Radiation Heat Transfer Modelling with Computational Fluid Dynamics

Radiation Heat Transfer Modelling with Computational Fluid Dynamics

Yehuda Sinai

CRC Press
Taylor & Francis Group
Boca Raton London New York

CRC Press is an imprint of the
Taylor & Francis Group, an **informa** business

Cover image: Magnum Fire, Kaibab National Forest, USA. Reproduced from Flickr

First edition published 2022
by CRC Press
6000 Broken Sound Parkway NW, Suite 300, Boca Raton, FL 33487-2742

and by CRC Press
4 Park Square, Milton Park, Abingdon, Oxon, OX14 4RN

CRC Press is an imprint of Taylor & Francis Group, LLC

Library of Congress Cataloging-in-Publication Data
Names: Sinai, Yehuda L., author.
Title: Radiation heat transfer modelling with computational fluid dynamics/Yehuda L. Sinai.
Description: First edition. | Boca Raton : CRC Press, 2022. | Includes bibliographical references and index. | Summary: "Radiation Heat Transfer Modelling with Computational Fluid Dynamics serves as a reference for principles of thermal radiation and its modelling in computational fluid dynamics (CFD) simulations. Including strategies for combining CFD and thermal radiation, the book covers computational techniques for solving the Radiative Transfer Equation, the strengths and weaknesses thereof, boundary and initial conditions, and relevant guidelines. Describing the strategic planning of a typical project, it includes spectroscopic properties of gases, some particulates, and porous media. The book is intended for researchers and professionals who simulate problems that involve fluid flow and heat transfer with thermal radiation"-- Provided by publisher.
Identifiers: LCCN 2021050628 (print) | LCCN 2021050629 (ebook) | ISBN 9780367766115 (hardback) | ISBN 9780367767884 (paperback) | ISBN 9781003168560 (ebook)
Subjects: LCSH: Heat--Radiation and absorption. | Heat--Transmission. | Computational fluid dynamics.
Classification: LCC TJ260 .S577 2022 (print) | LCC TJ260 (ebook) | DDC 621.402/2--dc23/eng/20211123
LC record available at https://lccn.loc.gov/2021050628
LC ebook record available at https://lccn.loc.gov/2021050629

ISBN: 978-0-367-76611-5 (hbk)
ISBN: 978-0-367-76788-4 (pbk)
ISBN: 978-1-003-16856-0 (ebk)

DOI: 10.1201/9781003168560

Typeset in Times
by SPi Technologies India Pvt Ltd (Straive)

Dedication

'To the memory of my parents,
Dolly and Reuben.'

Contents

List of Figures

List of Tables

Preface

This book provides a high-level overview of thermal radiation, the physics thereof, and computational methods for calculating thermal radiation in a typical terrestrial engineering environment and in the context of approaches using computational fluid dynamics (CFD). The most relevant topic is heat transfer, but the material herein can sometimes be applied in other spheres too. It is expected that the book will be of particular value to analysts facing a broad spectrum of topics. It complements and does not compete with the existing excellent textbooks on thermal radiation on the one hand, and on CFD (including user manuals) on the other, and cites references for the reader looking for greater detail. As such, it is a short introductory overview, with a small selection of 'tasters' of some of the large quantity of detail which exists. The book begins with a very brief outline of CFD, followed by a description of the strategic planning of the typical CFD project involving radiation. Then, as part of the background material, it distils key information on broad physics.

Special attention is paid to the coupling of the radiation physics and modelling approaches to the CFD processes and strategies.

This book is aimed both at CFD practitioners (especially in engineering disciplines) who are newcomers to thermal radiation, and at workers familiar with radiation who are newcomers to CFD, although there is a greater emphasis here on the former group. As such, the book is pitched as a broad-based introduction before the reader turns to much greater details in CFD user manuals or the specialist textbooks. The book assumes readership with at least a basic undergraduate level knowledge of flow and heat transfer, although a more advanced background in these fields will be beneficial. The readers may be students, researchers or professionals.

After the introductory material on radiation, which can be skipped by those familiar with the fundamentals, the book reviews when and how radiation is coupled to the CFD process, models available for the spectroscopic properties of gases, some particulates, and porous media, and then moves on to the primary computational techniques used to predict the radiative field and heat fluxes. Guidance is provided on the pros and cons of the methodologies, and attention is paid to the balance between accuracy on the one hand, and computing time and costs on the other, together with a few words about Quality Assurance.

Several examples are provided demonstrating applications of CFD with thermal radiation. This is followed by a small selection of analytical solutions which can be used to verify the radiation solver being deployed by the CFD practitioner.

Acknowledgements

I wish to thank Professor D. F. Fletcher and Dr P. J. Stopford (deceased) for their helpful comments.

The material in Section 7.1 was provided by ANSYS, and the CFD work described therein was carried out by G. Krishnamoorty, M. Sami, S. Orsino and P. Nakod. Section 7.2 is based on public-domain material issued by NIST. Section 7.3 was provided by Dr G.C. Fraga. Section 7.4 was provided by Mentor, a Siemens business. These have all been received with thanks.

A few items in this book are taken from (Sinai *et al.*, 2016). That material was developed under contract to IHS Markit ESDU, is published in the ESDU Internal Flow Series and is used with permission.

Permissions:

Figure 4.7, Incropera, F.P., DeWitt, D.P., Bergman, T.L. & Lavine, A.S., 'Incropera's Principles of Heat and Mass Transfer', Global Edition, 2017. Copyright © The content provided in this textbook is based on Incropera, Bergman and Lavine's Fundamentals of Heat and Mass Transfer 8th Edition (2017). John Wiley & Sons Singapore Pte. Ltd.

Figure 5.11, Reproduced with permission of the American Association for Aerosol Research.

Figure 5.12, reprinted with permission from (Radney *et al.*, 2014). Copyright (2014) American Chemical Society. Original received from Dr Radney with thanks.

Figure 7.14 to Figure 7.16: © Copyright Siemens.

Disclaimer

It is the reader's responsibility to ensure that the assumptions, methodologies and results from any analysis which has been performed as a result of, or in conjunction with reading this book, are correct and appropriate. Neither HeatAndFlow Consultancy Ltd. nor the author can accept any liability for incorrect analysis.

Author

Yehuda Sinai qualified in mechanical engineering at Witwatersrand University (South Africa) in 1968. He later travelled to the United Kingdom, where he earned a PhD in aerodynamics at Cranfield University in 1975, on non-equilibrium molecular relaxation effects in sonic booms. After a spell in academe working on acoustics and two-phase fluid-structure vibrations, he spent 10 years at National Nuclear Corporation Ltd. (NNC) on mathematical modelling of thermal-hydraulics at nuclear power stations, including gas-cooled, water-cooled and breeder reactors. He joined AEA Technology in 1992, where he specialised in CFD modelling in the safety sphere and became CFD Project Manager for Fire, Safety, HVAC and Environment. This role continued after the acquisition of AEA's CFD operation by ANSYS Inc. in 2003. In July 2009, Yehuda set up his own company, HeatAndFlow Consultancy Ltd., offering general modelling with a focus on CFD tools. He has authored and co-authored approximately 60 scientific papers in archival journals and conference proceedings, and besides this book is the author of a chapter in a technical encyclopaedia.

List of Symbols

a	Sound speed (m s^{-1})
A	Area (m^2), or parameter in Antoine relation, Equation (J.1)
$\hat{A}$	Constant in linear boundary condition, Equation (5.1)
$\hat{A}_1$	Cauchy parameter in double glazing model, Equation (K.4)
$\hat{A}_2$	Cauchy parameter in double glazing model, Equation (K.4)
A_S	Surface area of conducting solid, (m^2)
B	Parameter in Antoine relation, Equation (J.1)
$\hat{B}$	Constant in linear boundary condition, Equation (5.1)
Bi	Biot number
Bo	Boltzmann number, Equation (4.55), also known as Thring number
c	Speed of light
c_o	Speed of light in vacuo
C	Parameter in Antoine relation, Equation (J.1)
C_1	Constant in Planck function, Equation (4.4)
C_2	Constant in Planck function, Equation (4.5)
$\hat{C}$	Constant in linear boundary condition, Equation (5.1)
C_f	Friction coefficient
C_p	Specific heat and constant pressure (J kg^{-1} K^{-1})
C_{TRI1}	Constant in Equation (4.68)
C_{TRI2}	Constant in Equation (4.68)
C_S	Smagorinsky constant, Equation (4.69)
C_μ	Turbulence constant, usually $= 0.09$
d	Particle diameter (m)
$\hat{d}$	Size parameter, Equation (4.34)
D	Particle diameter in a porous medium (m), Equation (5.26)
$\mathcal{D}$	Binary mass diffusivity for a binary system (m^2 s^{-1})
$\mathcal{D}_{ij}$	Multicomponent diffusivity for the pair i–j (m^2 s^{-1})
$D_{\mathbf{mon}}$	Diameter of soot primary monomer (m)
e_i	Emissive power of surface i, σT_i^4 (W m^{-2})
e_g	Gas (fluid) emissive power, σT_g^4 (W m^{-2})
E	Emissive power (W m^{-2})
E_3	Exponential integral, Equation (E.2)
E_a	Activation energy (J kg^{-1})
E_b	Blackbody emissive power, σT^4 (W m^{-2})
E_n	Exponential integral, Equation (E.2)
E_w	Wall blackbody emissive power (W m^{-2}).
f_v	Particle or soot volume fraction, Equation (5.30)
F	Porous media coefficient, Equation (5.26)
F_{ig}	View or exchange factor between surface i and a volume
F_{ij}	Geometrical view or exchange factor
Fo	Fourier number
Fr	Froude number

$\hat{F}_r$ Densimetric Froude number

g Gravitational acceleration (m s^{-2})

G Incident radiation/intensity, or irradiation (W m^{-2})

Gr Grashof number

h Heat transfer coefficient (W m^{-2} K^{-1})

h_i Enthalpy of species i (J kg^{-1})

h_m Mass transfer coefficient (m s^{-1})

h_w Fluid-side heat transfer coefficient at a boundary (W m^{-2} K^{-1}).

H Specific, static enthalpy (J kg^{-1})

$\hat{H}$ Total enthalpy, $H_o + k$ (turbulence kinetic energy J kg^{-1})

H_o Stagnation enthalpy, $H + 0.5|\vec{u}|^2$ (J kg^{-1})

h_{GL} Heat of vaporisation (J kg^{-1})

I Total intensity/radiance (W m^{-2} sr^{-1})

$\bar{I}$ Mean intensity (W m^{-2} sr^{-1})

I_m Intensity associated with the medium (W m^{-2} sr^{-1})

I_w Intensity associated with walls/boundaries (W m^{-2} sr^{-1})

I_λ Spectral intensity/radiance (W m^{-2} sr^{-1} µm^{-1})

j_H Colburn j-factor

J Radiosity (W m^{-2})

$\hat{J}_i$ Diffusive mass flux of species i (kg m^{-2} s^{-1})

k Thermal conductivity (W m^{-1} K^{-1})

k Absorptive index, Equation (4.10)

k Turbulence kinetic energy (m^2 s^{-2})

k_B Boltzmann constant, 1.3807×10^{-23} J K^{-1}

k_g Conductivity (molecular) of the fluid (W m^{-1} K^{-1})

k_R Radiation conductivity (W m^{-1} K^{-1}), Equation (5.26)

K Extinction coefficient $K_a + K_s$ (m^{-1})

K_a Absorption coefficient (m^{-1})

K_{aP} Total Planck-mean absorption coefficient, Equation (5.25)

Kn Knudsen number, Equation (2.1)

K_s Scattering coefficient (m^{-1})

Ku Kutateladze number 1/Ja

L Characteristic length of the fluid medium, or slab thickness (m)

Le Lewis number

L_{mb} Mean beam length (m, Section 4.3)

L_p Characteristic separation between particles (m)

$\hat{L}_p$ L_p/λ, Equation (4.35)

L_S Characteristic linear dimension of conducting solid, V_S/A_S (m)

m Complex index of refraction, Equation (4.10)

$\dot{m}_G$ Gas mass flow rate in a channel (kg s^{-1})

$\dot{m}_L$ Liquid mass flow rate in a channel (kg s^{-1})

M Pope's LES metric, Equation (2.24)

n Refractive index

$\tilde{n}$ Ratio of refractive indices, n_2/n_1, Equation (C.1)

N Particle size distribution (5.28), or number of bounding surfaces

N_c Number of chemical species in a multicomponent mixture

N_{mon} Number of primary monomers in a soot cluster, Equation (5.29)
N_R Radiation number, Equation (4.56)
N_{Th} Thring number, Equation (4.55), also known as Boltzmann number
Nu Nusselt number
$\tilde{p}$ Scattering phase function, Equation (4.41)
P Pressure (Pa)
Pe Peclet number
Pr Prandtl number
q Heat flux (W m^{-2})
q_E Incident radiative flux from the exterior (Appendix K)
q_I Incident radiative flux from the interior (Appendix K)
q_R Radiative heat flux (W m^{-2})
Q Energy flow or load (W), or extinction efficiency, Equation (5.28)
R Gas constant, $\mathcal{R}/W$ (J kg^{-1} K^{-1})
$\mathcal{R}$ Universal gas constant, 8.3143 kJ kg-mole^{-1} K^{-1}
R Reflection coefficient (Section 4.1.9 and Appendix B)
Ra Rayleigh number
Re Reynolds number
R_G Radius of gyration of soot cluster (m)
Ri Richardson number
R_{ij} Distance between surface elements i and j
$\hat{R}_n$ Reflection coefficient at normal incidence (Section 4.1.9 and Appendix B)
$\hat{s}$ Unit direction vector
s_r Steradian
S Strain rate scale (s^{-1})
Sc Schmidt number
Sh Sherwood number
S_R Radiative source term in the enthalpy equation, Equation (4.60)
St Stanton number
t Time (s)
T Temperature (K)
$\bar{T}$ Reynolds-averaged temperature (RANS), or filtered temperature (LES)
T' Temporal temperature fluctuation (RANS), or difference between instanta-
          neous temperature and filtered temperature (LES)
T'_D Weighted source temperature (K), Section 5.1, Equation (5.15)
T_{DC} Source temperature for convection (K), Section 5.1, Equation (5.15)
T_{DR} Source temperature for radiation (K), Section 5.1, Equation (5.15)
T_i Temperature of surface i (K)
T_g Gas/medium temperature (K)
T_m Mean temperature $= 0.5(T_g + T_w)$, or Equation (5.17), (K)
$\hat{T}_n$ Transmission coefficient
T_R Mean radiant temperature, Equation (4.40), or radiant temperature displace-
          ment, Equation (K.2)
T_w Wall temperature (K)
T_S Equivalent or fictitious source temperature, Equation (K.1)
T_* Friction temperature, Equation (2.15)

$\vec{u}$	Flow vector (m s^{-1})
$\overline{u_i}$	Reynolds-averaged (RANS) or spatially filtered (LES) velocity component (m s^{-1})
u_τ	Friction velocity (m s^{-1}), Equation (2.10)
u_+	Dimensionless friction velocity, Equation (2.11)
u_*	Dimensionless friction velocity, Equation (2.12)
U	Characteristic speed (m s^{-1})
V	Domain volume (m^3)
V_S	Volume of conducting solid (m^3)
W	Molecular weight (kg per kg-mole)
x,y,z	Cartesian co-ordinates
y	Distance normal to a wall (m)
y_+	Dimensionless normal distance to a wall (m), Equation (2.13)
y_*	Dimensionless normal distance to a wall (m), Equation (2.14)

GREEK

α	Absorptivity, Equation (4.8)
$\hat{\alpha}$	Parameter in slab solution, Equation (E.10)
$\hat{\beta}$	Parameter in slab solution, Equation (E.10)
Γ	Gamma function
δ	Boundary layer thickness, m
Δ	Difference
Δ	Filter size in LES (m)
$\hat{\Delta}$	Parameter in slab solution, Equation (E.9)
$\Delta\rho$	Prescribed density difference, for example between jet inlet and the ambient (kg m^{-3})
ε	Emissivity
ε	Turbulence dissipation rate (m^2 s^{-3})
ε_g	Gas/medium emissivity
η	Wavenumber $1/\lambda$ (m^{-1}, or typically cm^{-1})
θ	Zenith angle (Figure D.1)
λ	Wavelength (m, or typically μm)
λ	Molecular thermal diffusivity, $k/\rho C_p$ (m^2 s^{-1})
$\hat{\lambda}$	Mean free path, or wavelength
μ	Dynamic viscosity (kg m^{-1} s^{-1}), or direction cosines
ν	Kinematic viscosity/diffusivity, μ/ρ (m^2 s^{-1})
$\tilde{v}$	Parameter in optically thick limit of Williams theory, Equation (G.9)
ρ	Reflectivity, or density (kg m^{-3})
ρ_o	Reference (usually ambient) density (kg m^{-3})
σ	Stefan-Boltzmann constant, Equation (4.4)
$\hat{\sigma}$	Surface tension (N m^{-1})
σ_ϕ	Prandtl number for dependent variable ϕ
τ	Opacity/optical depth $= KL$, or transmissivity
$\hat{\tau}$	Shear stress component (N m^{-2})
$\tilde{\tau}$	Stress tensor (N m^{-2})

τ	Optical distance, $K_a x$
τ_g	Opacity/optical depth
τ_L	$K_a L$
τ_s	Optical depth
τ_w	Wall shear stress (N m^{-2})
ϕ	Any of the dependent variables being computer by the flow solver
ϕ, φ	Azimuthal angle (Figure D.1)
ϕ_i	Weighting in S_N model, Equation (5.33), or radiant source per unit area of a slab (Appendix K)
ω	Scattering albedo, Equation (4.43)
ω	Turbulence frequency (s^{-1})
Ω	Solid angle (sr)
$\hat{\Omega}$	Angular speed, rad s^{-1}
$\vec{\Omega}$	Direction

SUBSCRIPTS AND SUPERSCRIPTS

$\overline{}$	Overbar: Time mean (RANS) or filtered (LES)
b	Blackbody
B	Pertaining to location B, Section 5.1
BC	Pertaining to region between B and C, Section 5.1
c	Convective
C	Pertaining to location C, Section 5.1
C	Collimated
C	Convective heat, or natural convection
CD	Pertaining to region between C and D, Section 5.1
D	Pertaining to location D, Section 5.1
D	Diffuse
E	Exterior
eff	Effective
F	Forced convection
g	Gas, or medium
G	Gas in a multiphase mixture
ℓ	Laminar/molecular
L	Liquid in a multiphase mixture
m	Associated with the medium
r	Radiative
R	Radiative
s	Pertaining to a solid
sat	At saturation
t	Turbulent
w	At or associated with a wall
λ	Dependent on wavelength
∞	Undisturbed ambient far from the body/surface generating the flow or disturbance

Acronyms

BC	Boundary condition
CFD	Computational fluid dynamics
CHT	Conjugate heat transfer
CK	Correlated k-distribution
DNS	Direct numerical simulation
DO	Discrete ordinates
DT	Discrete transfer
DTRM	Discrete transfer radiation model
EM	Electromagnetic
FE	Finite element
FS	Full spectrum
FSI	Fluid-structure interaction
FSK	Full-spectrum k-distribution
FVM	Finite volume method
GG	Grey gas
GWB	Grey wide band
GWSGG	Grey-Weighted Sum of Grey Gases
HTC	Heat transfer coefficient
HVAC	Heating, ventilation and air conditioning
IDA	Improved differential approximation
IR	Infrared
KD	k-Distribution
LBL	Line-by-line
LES	Large-eddy simulation
MDA	Modified differential approximation
MODIS	NASA's moderate resolution imaging spectroradiometer
NBM	Narrow band model
NIST	National Institute of Standards and Technology, USA
PDE	Partial differential equation
RANS	Reynolds-averaged Navier Stokes
RTE	Radiative transfer equation
SLW	Spectral-line-based WSGG
SNB	Statistical narrow band
TRI	Turbulence-radiation interaction
URANS	Unsteady Reynolds-averaged Navier stokes
UV	Ultraviolet
WBM	Wide-band model
WGGM	Weighted grey gas model
WSGG	Weighted sum of grey gases

1 Introduction

Thermal radiation is one of three recognised modes of heat transfer, the other two being conduction and convection. Phase change and the associated latent heat is sometimes described as a mode too. In some cases the contribution of radiation to the total heat transfer is small or negligible compared with the other modes, but in other cases it is comparable with or larger than the other modes. Several examples in which radiation plays a role follow:

Process and energy

1. Solar energy. Direct sunlight is exploited in solar heating panels and solar farms, and it is necessary to understand the interaction of radiation with the atmosphere and the device.
2. Glass manufacturing. Heat transfer in glass furnaces, which operate at high temperatures. Burners above the glass maintain its molten state, required during the manufacture.
3. Furnaces and boilers at power stations. Heat is used to generate steam in the boiler, and radiation is typically the dominant contributor thereto. Decarbonisation is of course leading to a decline of power generation employing fossil fuels but they are widespread at present.

Built environment

1. Thermal dynamics of buildings. The building structure is affected by radiation arriving from the external environment and from internal heat sources, and such matters affect indoor air quality, comfort and emissions.
2. Thermal comfort, affected by direct sunlight as well as diffuse radiation in the so-called shortwave and longwave ranges.

Transport

1. Combustion chambers in turbojets. Combustion is used to produce the high-energy gases which drive the turbine at the rear of the turbojet. The turbine drives the compressor and generates thrust, and thermal radiation is a major side-effect of the process, needing careful design of the components.

Environment

1. Meteorology and climatology. Interactions of the atmosphere with incoming solar radiation as well as with radiation reflected and emitted by the ground, the sea and the atmosphere itself. This includes the interaction with

DOI: 10.1201/9781003168560-1

gases as well as particulates such as water droplets (e.g. in clouds), ice crystals, soot, and dust. This topic is of course key in relation to climate change.
2. Remote sensing. Measurement of entities at the ground and in the atmosphere by terrestrial and satellite-born instruments.

Safety

1. Fire. Radiation is often the dominant factor inside the flame with implications for any structures engulfed by it. Radiation is also relevant at locations remote from the flame. Radiation plays a key role in the coupling between the flame and the fire (fuel) source for certain types of fire. Fire simulation is undertaken during design or emergency planning, or as part of forensic investigation of an accident.
2. Gas flares at oil and gas facilities (e.g. oil and gas platforms). These need to be designed with acceptable radiation at locations where workers may be present.
3. Explosions. Radiation may play a role in the rapid initial phase, but the slower fireball which follows it can also cause much harm and damage.

Medicine

1. Biomedical imaging. Measurement of electromagnetic waves passing through the body, leading to diagnostic tools which are more informative than X-rays.

Thermal radiation is an electromagnetic phenomenon, and is special and complex; special because it is capable of delivering energy remotely (unlike the other modes of heat transfer), and complex because it can depend not only on spatial co-ordinates and time, but also on direction, the spectroscopic properties of the medium though which the waves are travelling, and the wavelength of the waves. The mechanisms which influence the evolution of radiation are absorption (in which energy is transferred from the radiation field to the medium), emission (in which energy is transferred from the medium into the radiation field), and scattering (which involves direction change of the radiation field due to inhomogeneities). Refraction occurs when the medium's refractive index varies. There is substantial excellent literature on general heat transfer and on radiative transfer. For general heat transfer the reader is referred to typical examples such as (Simonson, 1988; Rohsenow, Hartnett and Cho, 1998; Holman, 2018; Lienhard IV and Lienhard V, 2012; Böckh and Wetzel, 2012; Incropera, DeWitt, Bergman and Lavine, 2017), and more, including some classics specifically on thermal radiation which are out of print (Chandrasekhar, 1960, 2013; Busbridge, 1960; Hottel and Sarofim, 1967; Sparrow and Cess, 1978; Ozisik, 1985; Brewster, 1992; Viskanta, 2008), as well as what can be described as the two leading contemporary books in this area (Modest, 2021; Howell, Menguc, Daun and Siegel, 2021). Important supplementary material, such a references, errata, and coding are provided for both of the latter books: Modest at http://booksite.elsevier. com/9780123869449/ and Howell at http://www.thermalradiation.net/. Climatology and atmospheric physics is of course an important topical issue, e.g. (Liou, 2002; Jacobson, 2005; Bohren and Clothiaux, 2006; Salby, 2012; Kondrat'Yev, 1965), and a little more will be said about multiphase aspects below. This book will generally avoid citing these references repeatedly, albeit with the occasional exception.

Reference has already been made above to waves: Thermal radiation may be represented both in terms of electromagnetic waves or bundles of energy called photons (associated with quantum theory), and the material in this book cites both approaches, as is typical in the literature, which refers to the 'wave-particle duality'. Normally the behaviour of solids, liquids, and interfaces is described more effectively by wave theory, whereas that of gases by quantum theory.

It is important to distinguish immediately between radiative transfer in non-participating (i.e. radiatively transparent) media on the one hand, and in participating media on the other. In a transparent medium, the opacity (defined in Section 4.2) is zero or very small, so that there is negligible absorption, emission, or scattering of radiation, and significant exchange only occurs between surfaces; this is known as surface-to-surface radiation, and the medium is normally described as optically thin or transparent. If the opacity is $O(1)$, i.e. of the order of 1, or large, the medium is said to be participating; in such a medium absorption and emission, as well as scattering, may exist in various proportions. Generally, poor conductors, known as dielectrics, enable long-range propagation, whereas good conductors are opaque. The general situation is governed by a complex (integro-differential) equation known as the radiative transfer equation (abbreviated here as RTE and described in Section 4.2), for which analytical solutions are rare, but several numerical algorithms are available in software packages. Simple engineering methods, using electrical circuit analogies or equivalents, are sometimes employed too, even for participating media (rarely nowadays), but the general computational algorithms are much easier to use, and are more flexible, albeit with a central processing unit (CPU) penalty. It should be noted that the electrical analogy, strictly speaking, is not a solution to the problem but is rather a reformulation. It uses surface-to-surface, surface-to-gas and gas-to-gas view factors, which require evaluation, either from existing tables or from solutions of the RTE. If the opacity is large, the medium is described as optically thick (and opaque in the limit); in such circumstances the radiative process is dominated by exchange between surfaces and the medium if variations in the medium are small, and the so-called 'diffusion approximations' are valid. If variations within the medium are not small, then exchanges within the medium are important too, but the larger the opacity, the smaller the distances over which such exchanges are occurring. Having said all this, it should be remembered that modern radiation solvers are capable of analysing the complete range of opacities, albeit at a non-negligible computational costs in some situations.

The applications quoted above illustrate the very diverse spheres in which radiation plays a role. Given the focus of this book on radiation in the context of CFD, it is thermal radiation which is the most relevant here. Magnetohydrodynamics (MHD) and relativistic phenomena are specialist fields which are beyond the scope of this book.

Another sphere which is beyond the scope of this book is the growing field of nanotechnology, involved for example in the behaviour of composites and microelectronics. If separations between bodies are smaller than the dominant electromagnetic wavelength, then near-field and coherence issues are introduced, heat fluxes can increase by orders of magnitude, the radiative transfer equation approach (Section 4.2) breaks down, polarisation becomes non-random and important (the waves are coherent), and the full Maxwell equations are required for analysis. The reader is referred to (Modest, 2021; Howell et al., 2021) and current literature for details. The current book considers the more common situations, which are in the far field, with separations between bodies larger than the dominant electromagnetic wavelength, and with incoherent radiation.

On occasion, use is made in this book of the term 'fluid'. This term refers to a gas or liquid. Molten solids are modelled as fluids too, and CFD can address non-Newtonian flows.

As far as CFD is concerned, an opaque object will simply manifest as a boundary condition for the radiation solver outside the solid (e.g. in terms of a surface emissivity). The CFD model may, however, include the interior of the opaque object, for example, by using conjugate heat transfer (CHT) functionality. Such a utility involves a mesh inside the opaque object, with conduction accounted for (as an example), and volumetric thermal radiation usually excluded. Radiation in semi-transparent solids can be modelled as stationary fluids, or in the case of some CFD packages, explicitly as non-opaque solids; that pertains to glass.

A comment is made here about the magnitude of radiative fluxes compared with the contributions of other modes of heat transfer. Generalisations should be treated with caution, but allowing for that caveat, in forced convection regimes temperatures need to be relatively high for radiation to be significant compared with the other heat transfer modes, whereas in natural or mixed convection regimes radiation can be of the same order as convection at room temperatures. Since radiation scales on the fourth power of temperature, its contribution increases rapidly with temperature, but as an illustration of the unwise assumption that radiation is always unimportant at ambient temperatures, consider the natural convection regime in the air at a nominal 20°C say, at large Rayleigh number (in the turbulent domain). Using correlations, e.g. (Incropera et al., 2017), the convective heat transfer coefficient is estimated to be $1.56\Delta T^{1/3}$, where ΔT is the temperature difference between a surface and the bulk fluid adjacent to it (see Sections 4.3 and 5.1.1). For temperature differences of 1 and 10 Celsius, this yields 1.56 and 3.36 W m^{-2} K^{-1}, respectively. Turning now to radiative exchange at the same surface with another surface (made of similar material) facing it, the radiative heat transfer coefficient can be estimated to be about 5.4ε, where ε is the surface emissivity (Section 4.1.2). Thus, the convective and radiative fluxes are comparable unless the surface emissivity is very small.

After the introduction, this book proceeds to describe the principal models available for spectroscopic properties, as well as computational techniques for solving for the radiative field given the local properties. It then continues to compare the models, and provides some examples which involve CFD and radiation.

Regarding the range of applicability of the material in this book, much has general applicability, but the emphasis is on typical engineering scenarios and conditions and on heat transfer, with parts of the book focused on a temperature range from ambient to 2000°C, typical say of phenomena from scenarios under ambient conditions to turbojet combustion chambers. Tools for analysing radiative transfer are of course used in areas other than heat transfer too, for example, lighting, optics, remote sensing, nuclear shielding, biophysics and astrophysics. The latter is a discipline in which much of the early pioneering work on radiation was formulated.

Finally, the author wishes to point out that whilst this book focuses on theoretical (computational) approaches, great debt is of course owed to the many experiments which have and continue to be undertaken, and which provide insight into the phenomena, key data, and avenues for validation of theory. Some analytical results are quoted in this book in order to explain and highlight concepts, and to provide some benchmarks against which the computational methods may be tested. Benchmarks are either exact analytical solutions or highly accurate analytical or numerical solutions.

This book aims to provide an overview in a generic fashion, but it will on occasion dip into specific applications. Also, since the material relates to radiation modelling in the context of CFD, relevant CFD aspects are discussed, but great detail of the substantial general topic are avoided, and the interested reader can turn to the many sources in the literature, e.g. (Fletcher, 1991; Roache, 1998; Ferziger, Perić and Street, 2002; Hirsch, 2007; Versteeg and Malalasekera, 2007; Andersson et al., 2011; Pletcher, Tannehill and Anderson, 2012; Tu, Yeoh and Liu, 2018).

Indeed, as already explained, this book addresses the gap between CFD on the one hand, and radiative heat transfer on the other, but it is tilted somewhat towards the CFD practitioner wishing to add thermal radiation to the predictive CFD setup.

After the Introduction, the book begins with a brief outline of CFD, and a summary of a typical process for setting up a CFD model. The bulk of the remainder consists of background material (Chapter 4) on the one hand, and modelling techniques and advice (Chapter 5 and Chapter 6) on the other. Section 5.3.5 provides guidance on models of the medium's spectroscopic properties, and Section 5.4.3 on radiation solvers. Readers experienced in radiation may skip Chapter 4. Chapter 7 offers two examples. Generally, the book's core tends to be descriptive in nature, supported by a little mathematics, and some deeper mathematical details are provided in the appendices. One exception to this is the discussion of the basic mathematical characteristics of the radiative transfer equation in Section 4.2, due the vital importance of that equation and the associated concepts.

2 A Brief Outline of CFD

CFD is a major topic, arguably on a larger scale than radiation, because it is a tool used in a prodigious array of sectors covering a very wide range of physics.

This book aims at the gap between computational fluid dynamics on the one hand and thermal radiation on the other. However, as already explained above, greater emphasis is placed here on the radiation issues than on CFD per se, and this section will therefore be very brief. Nevertheless, for the expert on radiation modelling unfamiliar with CFD, it provides a broad-based outline with plentiful pointers to the relevant literature. Major advances have of course been achieved in this field since the middle of the 20th century, and the unfamiliar reader is initially referred to the substantial literature, on fluid dynamics generally, although just a selection is listed here, which cannot not do justice to the discipline and its many sub-topics, e.g. (Townsend, 1980; Streeter and Wylie, 1983; Landau and Lifshitz, 1987; Lamb, 1994; Batchelor, 2000; Mathieu and Scott, 2000; Baukal Jr, Gershtein and Li, 2000; Pope, 2000; Leschziner and Drikakis, 2002; Lesieur, Métais and Comte, 2005; Poinsot and Veynante, 2005; Tavoularis, 2005; Bird, Stewart and Lightfoot, 2006; Garnier, Adams and Sagaut, 2009; Young, Munson, Okiishi and Huebsch, 2010; Cushman-Roisin and Beckers, 2011; Hanjalić and Launder, 2011; Tritton, 2012; Morrison, 2013; Argyropoulos and Markatos, 2015; Johnson, 2016; Kuerten, 2016; Rodi, 2017; Goldstein, 2017; Nakayama, 2018; Elger, LeBret, Crowe and Roberson, 2020; Fox, McDonald and Mitchell, 2020), and on CFD techniques in particular e.g. (Fletcher, 1991; Roache, 1998; Ferziger, Perić and Street, 2002; Hirsch, 2007; Versteeg and Malalasekera, 2007; Andersson *et al.*, 2011; Pletcher, Tannehill and Anderson, 2012; Blazek, 2015; Tu, Yeoh and Liu, 2018). These lists mostly cite books, but there are of course many journal papers, reviews and monographs too. Regarding the first citations, on general fluid dynamics, there are too many subtopics to list here, and just a few are included, such a turbulence, combustion, geophysical flows and measurement techniques, the latter to ensure that the theoretician sustains contact with reality.

A few examples of topics in which CFD has been deployed are external flows over aircraft, spacecraft, ships, sailing boats, wind turbines, buildings, bridges, people (including swimmers and cyclists), insects and fish, internal flows in chemical plant, turbojets, steam turbines, blood vessels, lungs, food manufacturing plant, refrigerators, as well as buildings, aircraft and ships during normal ventilation and during accidents such as fire, and environmental flows such as meteorology, oceanography, and environmental pollutant dispersion, to name but a few.

DOI: 10.1201/9781003168560-2

2.1 PRELIMINARIES

Before discussing CFD, two categorisations need to be highlighted here. The first distinguishes between continuum and non-continuum regimes. That is relevant from the viewpoints of both flow and radiation, because of topics such as (and not confined to) astronautics and rarefied gas dynamics, as well as the ongoing and growing activities in nanotechnology (not all involving radiation though), for example in renewable energy, electro-mechanical systems, drug design, nano-medicine. This impacts theoretical approaches to both fluid flow and radiative transfer. Generally, the literature refers to

- Nano-scales, measured in nanometres, e.g. nanofluidics.
- Micro-scales, measured in micrometres, e.g. micro-turbines.
- Macro-scales, measured in metres, e.g. everyday machinery.

The flow medium is modelled as a continuum if the characteristic length scale is much larger than the spatial scale of small irregularities. The classical measure for a gas, where the irregularity is the atom or molecule, is the Knudsen number:

$$K_n = \frac{\hat{\lambda}}{L} \qquad (2.1)$$

where $\hat{\lambda}$ is the atomic or molecular mean free path for gas (and a characteristic molecular separation for liquids), e.g. (Chapman and Cowling, 1990; Tabor, 1991; Bird, Stewart and Lightfoot, 2006). L is a characteristic length scale of the problem being considered. For example, the characteristic length scale is the mean hydraulic diameter for axial flow along a conduit, and that is equal to the diameter if the conduit cross-section is circular. A reminder that the classical mean free path for a gas is

$$\hat{\lambda} = \frac{k_B T}{\sqrt{2}\pi d^2 P} \qquad (2.2)$$

where k_B is the Boltzmann constant, and d is an effective collision diameter. To avoid the ambiguity of molecular diameter this has also been expressed in terms of the molecular viscosity. Assuming a typical diameter of 0.3 nm, then for gases under standard temperature and pressure (STP) conditions the indicative values for the average separation between molecules and mean free path are 3 nm and 70 nm, respectively.

Traditionally, three regimes are identified (McAdams, 1954):

a) $K_n \lesssim 0.001$. The medium is regarded as a continuum, with well defined average bulk properties, and no-slip boundary conditions at walls, amenable to traditional computational techniques.

b) $0.001 \lesssim K_n \lesssim 2$. Transitional regime known as *slip flow*, that term applying to the situation at walls.

c) $2 \lesssim K_n$. Known as *molecular or free molecular flow*. Intermolecular collisions become less important than collisions with walls (if those are present).

As far as fluid flow and dynamics is concerned, this book is confined to the continuum regime (a). The reader interested in non-continuum scenarios is referred to (Kogan, 1973; Chapman and Cowling, 1990; Cercignani, 2000; Carlson, Roveda, Boyd and Candler, 2004; Josyula, Xu and Wadsworth, 2005; Shen, 2010; Karniadakis, Beskok and Aluru, 2006; Czerwinska, 2009; Markesteijn, 2011; Drikakis and Frank, 2015; Kavokine, Netz and Bocquet, 2021).

The reader should be aware of the difference between 'non-continuum' and 'inhomogeneous'. For typical fluid flow, the fluid may of course be inhomogeneous, for example due to density, temperature and chemical composition variations, different phase states (in multiphase flow), or the existence of porous media (dealt with below), but the inhomogeneities are assumed to be on a spatial scale which is much larger than the mean free path.

The second important categorisation differentiates between CFD techniques which require a computational mesh on the one hand, and meshless techniques on the other. Traditional CFD relies on a mesh as the basis for discretisation of the governing equations on a macroscopic scale, and during the early days of CFD the meshing required much human effort, and could be demanding on CPU power too. That was a primary driver behind efforts to develop meshless techniques, which offer a viable alternative not only to CFD but in other aspects of mechanics too (Garg and Pant, 2018). In fact, those methods overlap with the techniques developed for non-continuum and statistical mechanics, because they track events at clouds of points. There is a variety of approaches, with the so-called SPH (Smooth Particle Hydrodynamics) and Lattice Boltzmann methods being two examples. The methods are amenable to massive parallelisation. The functionality in terms of physics behind the state of traditional CFD but is under development. On the other hand, meshing technology has advanced greatly since the middle of the 20th century, as is elaborated below, for example due to the advent of unstructured meshes, interfaces with CAD, and parallelisation of the meshing calculation. Thus, the pressure to develop the meshless technology is not as great as it has been in the past, but the developments will undoubtedly continue. Be that as it may, this book is confined to meshed technology. The reader interested in meshless methods is referred to the literature, e.g. (Divo and Kassab, 2006; Katz, 2009; Pepper, Wang and Carrington, 2013; Hosain, Domínguez, Fdhila and Kyprianidis, 2019; Nee, 2020). The relevance of this issue to radiation is touched on in Section 5.6 below.

2.2 GOVERNING EQUATIONS

Given the continuum assumptions discussed above, fluid flow is usually formulated as a set of laws of nature, such as conservation of overall mass, chemical species, momentum, and energy, expressed as partial differential equations (PDEs). Ignoring turbulence initially, these are of the following generic form:

$$\frac{\partial(\rho\phi)}{\partial t} + \nabla \cdot \left(\rho \vec{u}\phi - \frac{\mu_\ell}{\sigma_{\phi\ell}} \nabla\phi \right) = S_\phi \tag{2.3}$$

Here ϕ represents the dependent variables, such as 1.0 for the continuity equation, velocity component for the momentum equations, mass fraction for a chemical species, static or stagnation enthalpy for the energy equation, and volume fraction for each phase in a multiphase simulation. ρ is the fluid (mixture) density, $\vec{u}$ is the mass-averaged velocity vector, σ_ϕ is the Prandtl number for variable ϕ, and the subscript ℓ refers to the molecular property, essentially at laminar conditions. The Prandtl number for the momentum equations equals 1. Details of all the source terms (S_ϕ) in all these equations will be omitted here apart from that for the momentum and energy equations (the latter is of particular interest in this book because of the focus on heat transfer), and they are of course available in the literature. The momentum equation is being highlighted here to aid the outline of some aspects of turbulence discussed below. It is

$$\frac{\partial(\rho\vec{u})}{\partial t}+\nabla\cdot(\rho\vec{u}\vec{u})=-\nabla p+\nabla\cdot\vec{\tau}+\rho\vec{g}+\vec{F} \tag{2.4}$$

where $\vec{g}$ is the gravitational vector, $\vec{F}$ is a collection of other body forces such as interfacial drag in multiphase flow or magnetic forces in magnetohydrodynamics. $\vec{\tau}$ is the stress tensor, which for Newtonian fluids is given by

$$\vec{\tau} = \mu\left[\nabla\vec{u}+(\nabla\vec{u})^T -\frac{2}{3}\vec{\delta}\,\nabla\cdot\vec{u}\right] \tag{2.5}$$

where $\vec{\delta}$ is the Kronecker Delta, controlling the normal (diagonal) component of the stress, and the superscript T denotes the transpose. The energy equation can be expressed in several ways, for example in terms of the static enthalpy, the stagnation enthalpy, internal energy, or the (translational) temperature. In terms of the stagnation enthalpy for a single-phase multicomponent mixture:

$$\frac{\partial(\rho H_o)}{\partial t}+\nabla\cdot(\rho\vec{u}H_o)=\frac{\partial P}{\partial t}+\nabla\cdot\left(k\Delta T+\sum_{i=1}^{N_c}\hat{h}_i\hat{J}_i +\vec{u}\cdot\vec{\tau}\right)+\hat{S}_h \tag{2.6}$$

where $\vec{\tau}$ is the stress tensor, N_c is the number of species, $\hat{h}_i$ is the stagnation enthalpy of species i weighted by its mass fraction, and $\hat{J}_i$ is the mass diffusive flux of species i. For reference, the literature often makes use of the substantive/advective derivative:

$$\frac{D}{Dt}=\frac{\partial}{\partial t}+\vec{u}\cdot\nabla \tag{2.7}$$

The modelling of diffusion involving multiple species can be complicated, often with idealisations based on binary Fick diffusion coefficients, but turbulence helps in this case because turbulent mixing generally overwhelms molecular diffusion (unless the Reynolds number is moderate or low). The third term in the braces, involving the stress tensor, is the viscous dissipation which is often ignored, but with exceptions such as high-speed flow, at supersonic and hypersonic speeds. $\hat{S}_h$ represents other volumetric heat sources, in W/m^3. For example, if the mixture is reacting, then the source is positive for exothermic reactions, and negative for endothermic reactions. In the presence of thermal radiation, positive and negative sources exist in general, dealt with in detail in Section 4.4.

A key point is that the above equations comprise a set of non-linear coupled PDEs, which are also coupled to algebraic equations of state, such as density in terms of pressure and temperature, and enthalpy in terms of temperature (and sometimes pressure). If the problem involves thermodynamic non-equilibrium, then these relations, and entities like speed of sound depend on at least a third dependent variable representing the status of departure from equilibrium (cf. Section 4.5).

The inviscid and adiabatic form of this set is known as the Euler equations, which are hyperbolic in character. The general equations are usually known as the Navier–Stokes equations and are parabolic in character.

The non-linearity in these equations and their coupling has been a huge challenge, and unsurprisingly much (and indeed most) work was done initially using linear approximations, laying vital foundations and yielding powerful insight and results, e.g. (Lamb, 1994). Non-linear mathematical analyses were undertaken over many years, but the transformational advances in CFD and computing have naturally accelerated the inclusion of the non-linearities.

2.3　GEOMETRY AND MESHING

Whether or not the CFD software is mesh-based, the geometry must be defined. Most packages offer their own geometry tools, varying from hexagonal only, with stepped boundaries, at one end of approaches, to arbitrary curved boundaries at the other end. Moreover, most packages offer interfaces with CAD formats, which speeds up the geometry creation dramatically if the geometry has already been created in the CAD package. In the built environment sector increasing use is being made of BIM (Building Information Modelling), which facilitates the exchange of geometry with building architects and engineers, although some tidying is usually required in order to satisfy CFD requirements.

There are many meshing algorithms too. In broad terms, they have evolved over the years from cartesian, to Body-Fitted-Coordinates (BFC), to multiblock BFC, to unstructured meshes, and dynamic mesh adaption. Unstructured meshing permits refinement where it is needed and coarsening where it is not, leading to major savings in computing time and cost. Figure 2.1 shows typical mesh elements.

Refinement at boundaries often involves mesh 'inflation', typically with prisms, allowing resolution of wall boundary layers. Figure 2.2 illustrates such features.

Mesh adaption is also available, both in a static sense for refining the mesh at boundaries, and in a dynamic sense, allowing refinement in the bulk during the simulation, for example at mixing layers and shock waves where gradients are large, or for transient movement of objects. The latter can involve a prescribed motion or a full, dynamic, transient FSI (Fluid-Structure Interaction) such as wing flutter whereby the solid's gross movement and deformation are part of the solution.

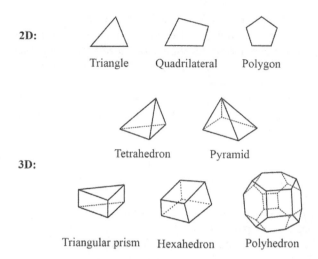

2D:

Triangle   Quadrilateral   Polygon

3D:

Tetrahedron   Pyramid

Triangular prism   Hexahedron   Polyhedron

FIGURE 2.1   Typical mesh elements.

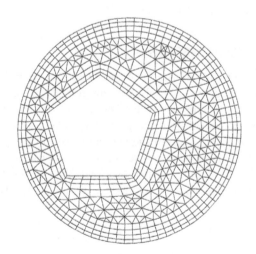

FIGURE 2.2   Illustration of inflated mesh between a pentagonal object offset inside a circular pipe, viewed perpendicular to the pipe axis. The mesh, which is tetrahedral in the bulk, has been deliberately coarsened here in order to aid visualisation of the mesh features. The number of inflation layers is reduced by the meshing algorithm at the narrower gap between the pentagon and the circle, and with a finer mesh this aspect would improve.

Domain decomposition divides the mesh into multiple regions for parallel processing, which greatly reduces run time. Parallelisation is a major area of activity.

2.4   MORE PHYSICS

A very wide range of physics is encountered by CFD practitioners. It is worth listing a few:

a) Steady or transient.
b) Incompressible.
c) Weakly compressible, and low Mach numbers.
d) Compressible, transonic, supersonic, hypersonic.
e) Non-buoyant, weakly buoyant, buoyant.
f) Laminar.
g) Transitional and turbulent.
h) Multiphase, with or without phase change (e.g. evaporation and condensation), multicomponent with inert species, miscible and immiscible fluids. Several multiphase regimes exist, not only particles dispersed in a background fluid. Three phases and multiple species may co-exist.
i) Thermal radiation, without and with participation (i.e. transparent and non-transparent/participating).
j) Chemical reactions.
k) Newtonian and non-Newtonian.
l) Thermodynamic non-equilibrium, such as molecular relaxation and ionisation.
m) Porous media.
n) Fluid-structure interactions; one-way and two-way interactions (mechanical, aeroelasticity, and/or thermal including Conjugate Heat Transfer).
o) Magnetohydrodynamics, plasmas.
p) Continuum and non-continuum, rarefied gas dynamics.

The cited textbooks cover most of these, and most of these topics are major disciplines in their own right. Thermal radiation is involved in many of these areas.

A few more words about multiphase flow are appropriate. That is a complex subject, and the interested reader is referred to the literature for details, e.g. (Hetsroni, 1982; Cheremisinoff, 1986; Brennen, 2005; Michaelides, Crowe and Schwarzkopf, 2016; Blunt, 2017; Yadigaroglu and Hewitt, 2017). The radiation literature dealing with multiphase aspects tends to focus on the dispersed regime, in which particles such as solid aerosols, liquid droplets or ice crystals, move within a carrier fluid (which may be a gas or a liquid); a little more is said about that in Section 4.4. However, it should be remembered that there are several other multiphase flow regimes. Flow regime maps exist in the literature, e.g. for flow inside horizontal, inclined or vertical tubes, based on experiments, e.g. (Baker, 1954; Mandhane, Gregory and Aziz, 1974), followed by unified theories (Taitel and Dukler, 1976; Barnea, Shoham, Taitel and Dukler, 1980; Barnea, 1987). The regimes for horizontal pipes are commonly identified as

a) Dispersed bubble
b) Stratified smooth
c) Stratified wavy
d) Intermittent-plug

e) Intermittent-slug
f) Annular
g) Dispersed (droplet/particle)

In vertical flow the *plug* term is replaced by *churn*. The last in the list is the one which most resembles the common multiphase topic tackled in the radiation literature. In many ways, some of the other regimes are more difficult to analyse, with or without radiation, especially the intermittent regimes (plug, slug, churn).

Mention should be made of another situation amenable to radiation modelling encountered in boiling, namely in the film regime and Leidenfrost phenomena. When the heat flux from the wall into the fluid or the wall temperature are large enough, a film of vapour (and non-condensibles if they are present) separates the wall from a liquid core; the latter may ultimately all be evaporated. Radiation across the vapour film is relevant. The converse situation applies too, with vapour from a gas mixture condensing onto a cold surface and forming distinct droplets, rivulets, or films. In that case the material at the wall tends to attenuate radiation in the infrared. Films are easier to model generally, and also from the radiation viewpoint.

Several multiphase models are offered by CFD codes. The most comprehensive but most expensive is called *Euler–Euler*, since it solves the Navier–Stokes equations for each phase, and addresses interfacial transfer. The regime which has received most attention with CFD tools is the dispersed domain, arguably because it is the regime most amenable to detailed modelling. Having said that, some codes are able to track individual phase interfaces in detail. Tools for the dispersed regime sometimes couple a Lagrangian analyses of 'superparticles', representing groups of particles (for example grouped by size), with an Eulerian analysis of the carrier fluid; this is cheaper computationally than the Eulerian–Eulerian approach. The particles can exchange mass and energy (including radiative heat) with the carrier fluid (and boundaries), and models are available for particle-wall interactions, including bounce. Population-dynamics models are offered by some codes. It is inevitable that with future advances in supercomputing the more challenging multiphase problems will be tackled in greater detail.

The next topic to be mentioned is dimensionless analysis, which provides a framework for understanding the problem, as well as designing experiments and the solution strategy. A fluid dynamics subset is included in this book in addition to those relating to radiation, and is given in Appendix A.

Of the many other issues, just turbulence will be mentioned here, very briefly. Many models are available, covered by the cited literature, and the broad categories available in CFD codes include, for example (listed in order of computational cost),

a) First-order two-equation RANS (Reynolds-Averaged Navier–Stokes), e.g. k-ε, k-omega, Realisable k-ε, k-ε RNG (Renormalisation Group), SST (Shear-Stress Transport), transition SST, SAS (Scale-Adaptive Simulation).
b) Second-order RANS, e.g. RSM (Reynolds Stress Model). Higher-order closures introduce more PDEs.
c) PDF (Probability Density Function) methods.
d) Hybrid, e.g. DES (Detached Eddy Simulation). RANS models are deployed close to walls and are dynamically and automatically blended with Large-Eddy Simulation (LES) in the bulk.

e) LES. Mesh sizes should be finer.

f) DNS (Direct Numerical Simulation). Much finer resolution of all scales of relevant motion, from Kolmogorov to integral scales.

In variable-density cases Favre averaging is often employed, whereby velocities are mass-weighted values. LES is becoming more common, although (a) and (b) are still in use. Hybrid methods are a compromise which reduces computing times and costs. DNS is the most refined approach, solving the transient Navier–Stokes equations directly without using turbulence closures, but is still too expensive (as of 2021) for practical large-scale engineering applications. The required number of floating-point operations per second (FLOPS) scales on $160Re^3$ (Pope, 2000), where Re is the Reynolds number based on the macroscopic length scale. Exascale computing (exceeding 10^{18} FLOPS) is being achieved as of 2021 on large mainframes, and growth of massive computing will undoubtedly continue.

Several classic fundamentally important scales related to turbulence will be mentioned here (Tennekes and Lumley, 1972; Pope, 2000). First, the Kolmogorov scales, at which the energy is dissipated, denoted by the subscript K:

$$\ell_K = \left(\frac{v^3}{\varepsilon}\right)^{1/4} ; t_K = \left(\frac{v}{\varepsilon}\right)^{1/2} ; u_K = (v\varepsilon)^{1/4} \tag{2.8}$$

Secondly, the large-eddy or integral scale is

$$\ell_I \sim \frac{k^{3/2}}{\varepsilon} \tag{2.9}$$

The constant multiplying it is typically $C_\infty^{3/4}$. Here v is the kinematic viscosity (m^2/s), k is the turbulence kinetic energy, and ε is the turbulence dissipation. The Taylor microscale, also known as inertial subrange, lies between these two extremes.

Two important classical entities, defined in several ways and affecting wall boundary conditions (inside the computational domain) are the familiar friction and associated velocities, usually denoted by u_+ or u_* or u_τ (sometimes the + appears as a superscript), and the normalised distance normal to the wall, essentially a Reynolds number, often called 'y plus' and denoted by y_+ or y_*. They are given by

$$u_\tau = \left(\frac{|\tau_w|}{\rho}\right)^{1/2} \tag{2.10}$$

$$u_+ = \frac{u}{u_\tau} \tag{2.11}$$

$$u_* = C_\mu^{1/4} k_y^{1/2} \tag{2.12}$$

$$y_+ = \frac{u_\tau y}{v_\ell} \tag{2.13}$$

$$y_* = \frac{u_* y}{v_\ell} \qquad (2.14)$$

where y is distance normal to the wall, and k_y is the turbulence kinetic energy at that distance. u_* is the usual entity employed in order to avoid numerical instabilities. Codes which impose a von-Kármán log-law at the wall require the first node to be in the log region, which lies in the approximate range $40 < y_* < 130$ (sometimes extended to $30 < y_* < 200$ and beyond). Most codes improve on this with a variety of techniques which are beyond this book's scope. A reminder that the classical inner mean velocity similitude involves three regions (Von Karman, 1931; Coles, 1956; Spalding, 1961; Dean, 1976; Whitfield, 1977; Swafford, 1983; Schlichting and Gersten, 2017; Tennekes and Lumley, 2018):

$y_* < 5$ Viscous sublayer
$5 \leq y_* \leq 30$ Buffer layer
$30 < y_* < 200$, $y/\delta < 0.3$ Log-law region (inertial sublayer)

where δ is the boundary layer thickness. Figures 2.3 and 2.4 show the classical (inner region) universal law of the wall for velocity and temperature respectively (the latter plot is for air), where the friction temperature is the usual

$$T_* = \frac{q_c}{\rho C_p u_*} \qquad (2.15)$$

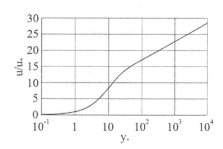

FIGURE 2.3   Universal law of the wall for velocity.

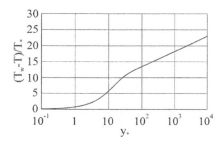

FIGURE 2.4   Universal law of the wall for temperature (in air).

All these are the familiar entities in the absence of interference by radiation; more is said about that in Section 4.4.1. Models are available in the outer (wake) regions of the boundary layer.

Omega-based turbulence models require the first node to be closer to the wall, at a distance comparable with or smaller than $y_* = 1$. Checking that y plus is small enough in the results, at all walls, is part of the quality assurance (QA) process.

Just two examples will be shown here of the generalisation of Equation (2.3), namely the first-order, eddy-viscosity form and the standard LES. Reynolds-averaging (i.e. time averaging) the PDEs in the RANS approach (denoted by an overbar) leads to equations for the mean quantities which look similar to the laminar equations, apart from the appearance of a stress tensor known as *Reynolds stress*. The second term on the right-hand side of Equation (2.3) becomes, in tensor notation,

$$\frac{\partial \tau'_{ij}}{\partial x_j} ; \tau'_{ij} = \tau_{ij} - \rho \overline{u_i u_j} \tag{2.16}$$

where the second term on the right-hand side is the Reynolds stress. Compressibility is taken into account by most codes by using Favre averaging. Analogous entities occur for other dependent variables. Representing those correlations, known as *closures*, has driven the distinct and major discipline of turbulence measurement and modelling. The eddy-viscosity approach is still the most popular in practical applications (as of 2021), and yields

$$\frac{\partial(\rho\phi)}{\partial t} + \nabla \cdot \left[\rho \vec{u}\phi - \left(\frac{\mu_\ell}{\sigma_{\phi\ell}} + \frac{\mu_t}{\sigma_{\phi t}} \right) \nabla\phi \right] = S_\phi \tag{2.17}$$

The subscript t denotes a turbulent entity. The sum involving the laminar (molecular) and turbulent viscosities defines an effective diffusivity, having dimensions m²/s:

$$v_{\phi eff} = \frac{1}{\rho} \left(\frac{\mu_\ell}{\sigma_{\phi\ell}} + \frac{\mu_t}{\sigma_{\phi t}} \right) \tag{2.18}$$

This approach is normally framed in terms of two PDEs. These two-equation models include one of the familiar model-dependent algebraic relations for the turbulent eddy-viscosity μ_t, in terms of dependent variables. For example, the classic expression for the k-ε model is

$$\mu_t = C_\mu \rho \frac{k^2}{\varepsilon} \tag{2.19}$$

where C_μ is a constant, typically 0.09. Another example is for SST (Menter, Kuntz and Langtry, 2003), in which the model is blended automatically between k-ω near the wall on the one hand, and k-ε in the bulk on the other; its eddy viscosity is

$$\mu_t = \frac{\rho \alpha_1 k}{\max\left(\alpha_1 \omega, SF_2\right)} \qquad (2.20)$$

where α_1 is a constant, and S is a strain rate (s^{-1}). F_2 is a smooth blending function which depends on distance normal to the wall, k, and ω. It is zero at the wall and 1.0 far from it. Another blending function appears in the turbulence differential equations, and governs the transition between the k-ω and k-ε forms.

The second technique outlined here very briefly is Large Eddy Simulation (LES), because its use is increasing in practical situations. Its modelling is simpler than RANS and is more comprehensive, at a higher computing cost; if the LES is set up as it should, then that cost is typically hundreds of times larger than RANS. The temptation to reduce that cost by coarsening the mesh carries a potential price, namely reduced accuracy.

Instead of time averaging, a spatial filtering is carried out. An example is the top-hat filter possessing a width of Δ, which in one dimension is

$$\bar{\phi} = \int_{-\infty}^{\infty} \phi(x')G(x')dx'; G(x) = \frac{1}{\Delta}\left[H(x+\Delta) - H(x-\Delta)\right] \qquad (2.21)$$

where $H(x)$ is the Heaviside function which equals 1 if $x > 0$ and zero when $x < 0$. Usually, the filter width is set to be $V^{1/3}$, where V is the local CFD cell volume.

The resulting equations are similar to Equation (2.19) and the RANS equations, which again leads to a sub-grid stress (abbreviated as SGS):

$$\tau_{ij}^{LES} = \overline{\rho u_i u_j} - \overline{\rho u_i u_j} \qquad (2.22)$$

The standard approach, named after Smagorinsky, employs the isotropic eddy-viscosity concept:

$$\tau_{ij} - \frac{1}{3}\delta_{ij}\tau_{kk} = -2v_t \overline{S_{ij}}; v_t = \left(C_s\Delta\right)^2 |S| \qquad (2.23)$$

where S_{ij} is the strain rate, and C_s is the Smagorinsky constant which is usually assigned a value of about 0.18. An important deviation from RANS occurs here, because the filter width is proportional to the computational cell size, and no further PDEs are needed in this case. However, mesh sizing is important, and can become a limiting factor especially in wall-bounded flows. The filter size needs to be larger than ℓ_K (Equation (2.8)) and smaller than ℓ_I (Equation (2.9)), see (Pope, 2000). Ideally a RANS simulation is performed first, in order to provide meshing guidance. Usually the aim is to capture about 80% of the turbulence kinetic energy; this is known as Pope's M criterion, which recommends $M \leq 0.2$, where

$$M = \frac{k_{SGS}}{k_{LES} + k_{SGS}} \qquad (2.24)$$

where k_{LES} is the local temporally averaged resolved kinetic energy (evaluated as a running time average of the LES solution), and k_{SGS} is the sub-grid kinetic energy, estimated by $(\nu_t/0.09\Delta)^2$. A number of techniques have been developed to cope with walls (particularly important in aerodynamics), and important developments over the years have also led to improvements using dynamic and adaptive LES, in which the filter size is not constant. For all these issues the reader is referred to the cited works.

In DNS the Kolmogorov scales are resolved and the exact laminar equations are solved. LES and DNS require transient boundary conditions at inlets and interfaces. For fully developed flows that can be achieved directly by imposing periodicity at the inlet and outlet, but for most situations that is not appropriate, and 'synthetic' turbulence can be specified in several ways using RANS simulations upstream of the domain, e.g. (Mathey, Cokljat, Bertoglio and Sergent, 2006) and user manuals.

2.5 NUMERICS

Early CFD used staggered grids for the pressure and three velocity components, in order to overcome the so-called checkerboarding phenomenon. Pressures were stored at cell centres and velocities at cell face centres. Collocated grids, in which all variables are stored at the same locations (nodes), are common nowadays.

Several discretisation schemes are in use. Figures 2.5 and 2.6 illustrate the spatial basis of some schemes. The dependent variables are stored at the dots, and the hatches denote the control volume. Fluxes are estimated at the control volume faces.

A variety of discretisation schemes are available. First-order is the crudest, and produces 'false' or 'numerical' diffusion which can overwhelm physical diffusion.

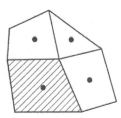

FIGURE 2.5 2-D sketch of cell-centred discretisation locations. Hatch denotes control volume, and • denotes cell centre.

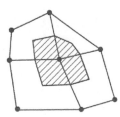

FIGURE 2.6 2-D sketch of a vertex-based discretisation locations. Hatch denotes control volume, and • denotes cell vertex/node (and control volume centre).

Second-order schemes are the norm nowadays, and some codes offer much higher-order resolution too.

Similarly, early solvers were of the segregated type, in which a subset of the governing equations is solver iteratively then coupled to other subsets, again iteratively. In contrast, some later developments have provided coupled solvers, in which some or all of the equations are solved at once. Coupled solvers require more memory but are generally more robust.

2.6   PROBLEM SIZE

As always, computing capability grows inexorably. Massive computing is reflected in the existence of organisations such as TOP500. Millions of cores and billions of grid points have being tackled and speed has increased too of course (cf. the discussion on turbulence above). Having said that, such supercomputing is certainly not employed by everyone, and everyday designs by many practitioners deploy resources which are far more modest than the top entries in TOP500.

This completes the cursory review, and much more is of course available in the cited literature and software user manuals.

3 Outline of a Typical Process for CFD Analysis with Radiation

A description will be given here of a representative project process for CFD with thermal radiation.

The first step defines the purpose and scope of the analysis, and formulates an understanding of how it fits in with the larger project of which it is usually a part. If the project is a commercial one, a balance usually needs to be struck between accuracy on the one hand, and both computing costs and run time on the other, and 'overkill' is unwise. Sometimes CFD is unwarranted altogether.

The second step is choice of software, and the primary distinction will be between bespoke software on the one hand (used for specific categories of problems), and general-purpose software. Bespoke versions of general-purpose software are sometimes also available from the provider, or may be created by the user if the necessary functionality is available.

A key factor is the level of quality assurance employed by the software provider as well as by the CFD practitioner. Chapter 6 touches on the latter, and the practitioner of course needs to be confident of adequate verification and validation of the software. Here 'verification' refers to comparison of the software predictions with exact mathematical or highly accurate numerical solutions for 'separate-effects' scenarios, in which only subsets of the physics phenomena are present, and 'validation' refers to comparison of the software predictions with physical experiments or full-scale monitoring of the product performance.

Once the software has been chosen, the typical process consists of the following:

a) Strategic planning.
   a. Spatial extent of domain.
   b. Overall scope of physics, including strategy for boundary conditions. Is the physics covered by what is included in the standard software? If not, can it be implemented by the user? If the software is commercial, meaning that the source is unavailable to the user, does the software provide user-routine functionality? User routines are modifiable templates in languages such as C++ or Fortran. Some packages provide scripting language utilities which are a halfway-house between no user intervention and full user routines. Depending on the specifics of the technical issue being tackled, this aspect may sometimes be implemented in freeware. Some providers operate in a different commercial fashion and offer freeware and the source coding.

DOI: 10.1201/9781003168560-3

    c. Any coupling with other software? If there is, how will that be achieved? Various levels of coupling exist, under the general heading of 'multi-physics'. Here are several examples:

       i.   Interactions between a fluid domain and a solid structure. Interactions may, for example, be thermal (heat exchange), mechanical (exchange of pressure-driven forces and accompanied movement of the solid). These come under the general heading of Fluid-Structure Interactions (FSI). Some CFD packages allow the modelling of solids as well as fluids at various levels of sophistication, the most common being Conjugate Heat Transfer (CHT), in which conduction and radiation are allowed in the solid. This is a powerful and useful facility. However, sometimes a more sophisticated structural package is appropriate, allowing greater detail and wider physics within the structure. Structural models usually employ finite-element techniques. Heat exchange between turbojet hot gases and turbine blades is an example of thermal coupling, as is heat exchange between tubes and the shell-side and tube-side fluids in a boiler or heat exchanger. If vibrations of the blades or tubes are to be predicted, then the setup involves both thermal and mechanical coupling.

      ii.   A detailed CFD model of an underground metro station coupled to a network model of the whole system of tunnels and train movement therein.

     iii.   A CFD model of a building's indoor environment coupled to a separate package which analyses glass windows with multiple panes accounting for the effects of the external environment, in order to provide a boundary condition to the CFD code.

b) Meshing, although some packages employ meshless technologies.

c) Pre-processing, which includes physics and numerical/computational settings. Examples, with more detail on radiation, given the subject of this book:

    a. Single-phase or multiphase?

    b. Turbulence model?

    c. Buoyancy?

    d. Incompressible or compressible?

    e. Inert or reacting?

    f. Radiation. Is it coupled to the flow? Transparent or participating? Grey or non-grey? Surface properties. Solution method, balancing cost against accuracy.

    g. CHT (Conjugate Heat Transfer)? In CHT the computational domain includes not only fluid regions but also solid heat-conducting regions which are fully coupled to the fluid regions. As mentioned, some packages can model radiation in the solid.

    h. Porous regions?

    i. Steady or transient?

      j. Boundary conditions (and initial conditions for transient cases).

      k. Settings for generation of tabular and graphical results output.

      l. Numerical and computational controls.

d) Run execution.

e) Post-processing.

f) Quality assurance.

g) Reporting.

4 Fundamentals of Thermal Radiation

This section introduces relevant fundamentals of the underlying physics and some associated terminology.

4.1 BASICS

Thermal radiation is a manifestation of electromagnetic waves and their interactions with the materials through which they propagate. These waves are made up of electrical transverse waves which are perpendicular to magnetic transverse waves. The implications for heat transfer are dependent on their direction of propagation, frequency, and energy content. The polarisation phenomenon can also play a role, especially in optics and material identification, but that is uncommon in engineering heat transfer scenarios. The literature on this subject is of course substantial; Chapter 1 has already cited sample literature, and in relation specifically to the physics of electromagnetism the reader is referred to (Hayt and Buck, 2019; Heald and Marion, 2012), two of many dealing with this topic.

4.1.1 ELECTROMAGNETIC SPECTRUM

Figure 4.1 shows the electromagnetic spectrum. Two main categories of radiation exist: Ionising and non-ionising radiation. The former is intense, at short wavelengths and high frequency, and is capable of disrupting chemical bonds in biological systems. Non-ionising radiation does not do so, and is the subject of this book. The human eye responds to radiation in the visible range of 0.38 to 0.7 microns, and solar radiation peaks within this range. The region straddling the solar peak is usually divided into three regimes: *Ultraviolet* (0.01 microns to 0.38 microns), *visible*, and *near-infrared* (also known as *shortwave*). In the literature, infrared is defined as 0.7 microns to 1000 microns (the start of the microwave range), and near-infrared as the range 0.78 to an upper limit cited in the literature to be between 2 µm and 5 µm. The literature also refers to the mid-infrared (5 to approx. 40 µm), and far-infrared beyond that. Some variations in terminology exist in the literature: For example, in atmospheric physics (e.g. (Salby, 2012)), and the built environment (e.g. (ASHRAE, 2017)) sectors the spectrum is often divided into two regions, with *'shortwave'* covering wavelengths less than about 4 µm, and *'longwave'* covering larger wavelengths, whereas in other sectors further divisions of the spectrum tend to be employed more frequently. The shortwave radiation is characteristic of emission by the sun (at a temperature of about 5800 K), and the longwave radiation is characteristic of emission by the Earth's surface and atmosphere (at typical ambient temperatures). Ultraviolet is itself divided into Extreme UV, Far UV, Middle UV or UV C, UV B, and UV A

DOI: 10.1201/9781003168560-4

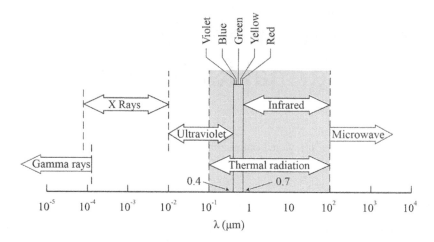

FIGURE 4.1 Part of the electromagnetic spectrum (Incropera *et al.*, 2017). Radio waves are beyond microwaves ($\geqslant 0.1$ m). In meteorology and built environment sectors the spectrum is also divided into two primary bands, called shortwave when $\leqslant 4$ µm, and longwave when $\geqslant 4$ µm.

(listed in increasing wavelength). Of the total solar energy, approximately 8% lies within the ultraviolet range, 47% within the visible range, and 45% within the near-infrared range.

Thermal radiation has been defined in the past as the electromagnetic waves which are generated by a medium due solely to its temperature, and is typically assumed to lie approximately in the range 0.1 to 100 microns.

If the radiative properties of a material are independent of wavelength, it is said to be '*grey*' (or '*gray*'), and if they do depend on wavelength or frequency, the material is said to be '*non-grey*' and its behaviour is '*spectral*'. The subscript 'λ' of any parameter often denotes spectral behaviour of that parameter.

Regarding the sun, the 'solar constant' refers to the extra-terrestrial incoming flux, which is not constant but varies by a few percent, principally with annual and 11-year cycles. Its characteristic value is 1370 W m^{-2}.

4.1.2 Black Bodies, Surface Behaviour, and Radiosity

A '*black*' body is an ideal emitter of thermal radiation, and its behaviour is measured in terms of 'emissive power'. The blackbody spectral emissive power for a body at absolute temperature T as a function of wavelength, is given by the classic Planck distribution:

$$E_{B\lambda} = \pi I_{B\lambda} = \frac{C_1}{n^2 \lambda^5 \left[\exp\left(\dfrac{C_2}{n\lambda T} \right) - 1 \right]} \tag{4.1}$$

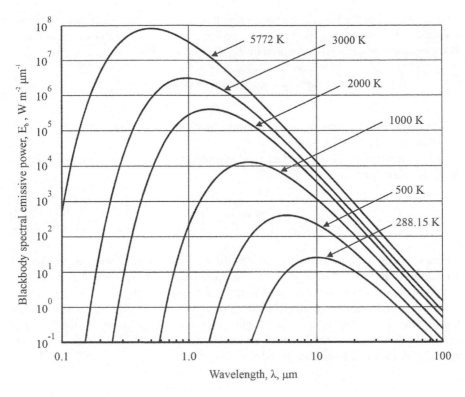

FIGURE 4.2 Planck spectral blackbody emissive power E_b, with shading over approximate visible range.

$$C_1 = 3.7418 \times 10^{-16} \, \text{W m}^2 = 3.7418 \times 10^8 \, \text{W} \, \mu m^4 \, \text{m}^{-2};$$
$$C_2 = 0.014388 \, \text{mK} \tag{4.2}$$

This is shown in Figure 4.2, where the emissive power is plotted as a function of wavelength at a number of specific temperatures, for a black surface adjacent to a transparent medium possessing a refractive index n equal to 1.0. For other values of n, the wavelength λ is replaced by $n\lambda$, and the emissive power is multiplied by n^3.

Note that the curve at room temperature peaks at about 10 μm, and that at $T = 5762$ K, the characteristic mean surface temperature of the sun, it peaks at about 0.5 μm, which is in the visible range. The locus of the peaks of the curves, a straight line known as Wien's displacement law for thermal radiation (expressed here in μm), is:

$$n\lambda_{\text{max}} = \frac{2897.6}{T} \tag{4.3}$$

where the numerator has units of μm·K, and T is in Kelvin.

The integral of the Planck distribution with respect to wavelength, over all wavelengths, is the 'total blackbody emissive power', often simply referred to as blackbody emissive power, and is denoted by E_b, defined as follows:

$$E_b = \sigma T^4; \quad \sigma = 5.6705 \times 10^{-8} \, \text{W m}^{-2} \, \text{K}^{-4} \tag{4.4}$$

σ is the Stefan–Boltzmann constant. An ideal black surface also absorbs all radiation incident upon it and reflects none of it.

The *total intensity I* at the surface (and at locations away from surfaces) is related to the *total emissive power* as follows:

$$I = \frac{E}{\pi} \tag{4.5}$$

where the denominator has the unit of steradians, the measure of solid angle. A similar relationship holds in the spectral sense, I_λ and E_λ replacing I and E respectively, having the units W m^{-2} sr^{-1} μm^{-1} and W m^{-2} μm^{-1} respectively.

Now consider real materials. It has been observed that usually the radiation emitted by materials is less than E_b (see Figure 4.3). A quantity known as the '*spectral emissivity*' is defined as the intensity emitted by a surface divided by the blackbody intensity at the same temperature and wavelength.

$$\epsilon\left(\lambda, T, \text{direction}\right) = \frac{I\left(\lambda, T, \text{direction}\right)}{I_b\left(\lambda, T\right)} \tag{4.6}$$

For a grey surface the emissivity is independent of wavelength, and the energy emitted by a surface per unit area is approximated by

$$E = \varepsilon \sigma T^4 \tag{4.7}$$

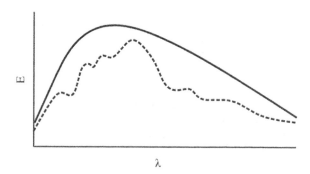

FIGURE 4.3 Wavelength/frequency dependence. Schematic of variation of emissive power of a real, non-grey surface with respect to wavelength, on a log–log plot. Solid line: Blackbody. Dotted line: Real surface.

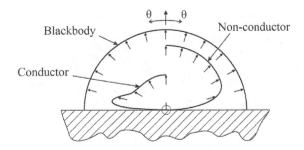

FIGURE 4.4 Directional dependence. Schematic polar plot of typical emitted intensities, or emissivities, at a given temperature and wavelength, at a black surface and two real surfaces: A conductor, and a non-conductor (dielectric). Symmetry is assumed about the normal.

where ε is the *grey surface emissivity*. In the literature the term *emittance* is used sometimes, usually to denote surfaces which are not optically smooth or contaminated (which is indeed the more common situation), but in practice both endings are in use, and in this book the '-ivity' ending is opted for.

Regarding direction, the emissivity usually only varies with the angle between the direction vector and the normal to the surface (Figure 4.4), although for some surfaces the behaviour will be more complex. For many materials it is common to assume independence of direction when using the more detailed computational models, and the surface is then known as 'diffuse' or 'Lambertian'. The assigned emissivity should usually be equated to the hemispherical average. As can be observed from Figure 4.4, the hemispherical emissivity is slightly smaller than the emissivity normal to the surface for non-conductors, but for conductors it is larger, to a greater extent than the reverse applies to non-conductors. Be that as it may, it is common practice to adopt the normal emissivity, which is easier to measure, even if this may change as computing power continues to increase.

Kirchoff's Law states that the surface emissivity is equal to the *surface absorptivity*, denoted here by α. When energy is incident upon a surface, some will be absorbed by the surface (and/or transmitted through it), and the remainder will be reflected. The *reflectivity* ρ at an opaque surface is given by

$$\rho = 1 - \alpha = 1 - \varepsilon \tag{4.8}$$

A surface which is rough on the scale of the wavelength of the radiation produces '*diffuse*' reflection, in which the reflections are in most of the hemisphere above the surface (Figure 4.5a), whereas a mirror-like surface is said to produce '*specular*' reflection in which the reflection is concentrated about the mirrored angle (Figure 4.5b). Real surfaces can be a combination of these two (Figure 4.6), and are modelled with both diffuse and spectral reflectivities, usually as a sum of a diffuse term and a specular term. A surface is said to be optically smooth when the size of its characteristic roughness elements are significantly smaller than the electromagnetic wavelength.

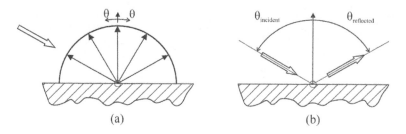

(a)   (b)

FIGURE 4.5   Surface reflection. The image illustrates incident radiation arriving at 60° to the normal. (a) is diffuse, (b) is specular, with the reflected angle equal to the angle of incidence.

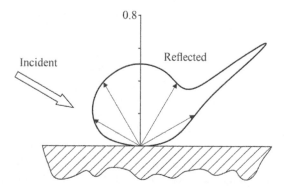

FIGURE 4.6   Schematic polar diagram of the reflectivity of a real, hybrid surface for a ray incident from the left at 60° to the normal.

Typical surface (normal) emissivities of a limited number of materials are shown in Figure 4.7, which is here to provide a guideline rather than specific values. It is emphasised here that emissivity can vary significantly not only from material to material and with wavelength, but also for a given material since it can be affected by its environment. A classic example is the rusting of metals such as steel, a process which affects the emissivity dramatically, as usefully demonstrated in the figure. In some cases the emissivity is affected by deposition of substances to which it is exposed, not necessarily involving rusting. This uncertainty should be born in mind when considering whether to embark on accurate but expensive techniques of radiation prediction. When dealing with materials possessing wide variations in emissivity (or other properties), especially if they are hard to quantify, the computational fluid dynamics (CFD) practitioner would be wise to perform a sensitivity analysis. Additional empirical emissivity data may be found in the cited textbooks and elsewhere, for example in (Singham, 1962).

A schematic of behaviour at a surface, or indeed an interface, is provided in Figure 4.8, without distinguishing between diffuse and specular reflections. The bold horizontal line represents a surface or boundary, and radiation is being considered in the region above the line. The surface emits into the domain, radiation is arriving at the surface from other sources in the domain (denoted by irradiation), with some of that being reflected and the remainder absorbed.

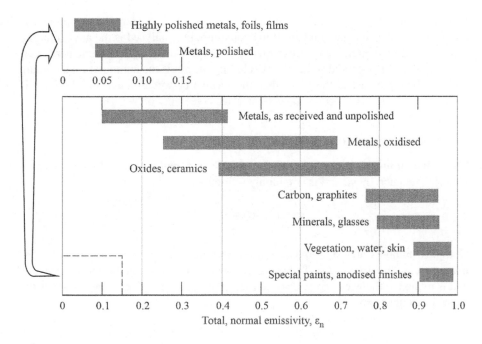

FIGURE 4.7 Some surface emissivity ranges. From (Incropera *et al.*, 2017).

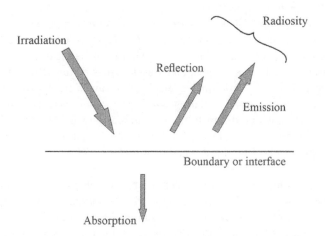

FIGURE 4.8 Schematic of behaviour at an opaque boundary or interface. In the former case the computational domain is above the horizontal line. The arrows do not necessarily imply specular or collimated behaviour.

The definition of *radiosity*, denoted by *J*, also shown in Figure 4.8, is

$$\text{Radiosity} = \text{emission} + \text{reflection} \tag{4.9}$$

Boundaries may of course be opaque or non-opaque. An example of the latter is the interface between glass and air. If the domain being analysed is the air, excluding the glass, then the interface is an external non-opaque boundary. The same pertains if the domain being analysed is the glass excluding the air. If the computational domain includes both the air and the glass, then the surface between the air and the glass is an interface. Refraction occurs at such interfaces (see Section 4.1.9).

4.1.3 SPEED OF LIGHT AND REFRACTIVE INDEX

The radiation literature refers to the dimensionless *complex index of refraction* of a material, denoted by *m*, and is commonly written as

$$m = n - ik; i = \sqrt{-1} \tag{4.10}$$

Complex numbers are commonly deployed in mathematical analysis of wave phenomena, e.g. (Tolstoy, 1973). The real component *n* is known as the *refractive index* and is related to the wave's phase speed *c* (i.e. the speed of a given phase, such as the crests or troughs), by

$$n = \frac{c_0}{c} \tag{4.11}$$

where c_0 is the speed of light in vacuo (2.998×10^8 m/s). For most gases *n* is very close to 1.0; for example, it is 1.00029 for air at room temperature in the visible range. For most dielectrics (i.e. poor electrical conductors), which include glass, *n* is fairly constant for a given material in the range between visible and mid-infrared, but tends to rise at smaller wavelengths, and varies across different materials approximately between 1.4 and 4 microns. Typical glass values lie approximately in the range 1.5 to 1.7, with a small variation within the visible range for a given glass. That small variation, with *n* diminishing with increasing wavelength (for most semi-transparent materials), is responsible for the classical separation of white light into a colour spectrum, with blue light (smaller wavelength) being refracted more than the red (larger wavelength). The frequency of the radiation wave does not change as the wave passes from one refractive index to another, whereas its wavelength does. In general the refractive index of media other than a vacuum varies with frequency, although the extent of that variation is diverse for different materials.

The parameter *k* is known as the *absorptive index*; it is large when the electrical conductivity is large, and is small when the material is a poor conductor. Thus electromagnetic waves tend to be attenuated rapidly in good conductors such as metals, and are weakly attenuated (i.e. propagate and are sustained over larger distances) in poor conductors (known as dielectrics or insulators). In a perfect dielectric *k* = 0. The complex index of refraction is often dependent on the wave's frequency (or wavelength), and common examples of that behaviour are glass, in which *k* is small in the visible range and large in the infrared, and atmospheric air, a multicomponent mixture which also displays the same general behaviour, albeit with many transmission 'windows'.

Values of the refractive index may be found in the cited literature and, for example, in (Hecht, 2017).

4.1.4 SHAPE FACTORS

Early engineering assessments of thermal radiation tended to focus on transparent situations, typically with radiant heat exchanged between diffuse surfaces bounding a transparent medium. That exchange is determined by (among other issues) the geometry of the surfaces and their orientations relative to one another. The *shape factor* is also referred to in the literature as *view factor, configuration factor* and *angle factor*, and is usually associated with radiation across transparent media. Having said that, it should be noted that such factors also arise in the classical literature on radiation across participating media, and in that context, relate to exchange between one surface and another surface, between one surface and a volume element, and between one volume element and another volume element.

Turning now to the transparent situation, consider the two small and black area elements dA_i and dA_j in Figure 4.9. The elemental view factor dF_{1-2} is the fraction of radiative energy leaving dA_1 which is intercepted by dA_2. Due to the diffuse assumption, we have

$$dF_{ij} = \frac{\cos\theta_i \cos\theta_j}{\pi R_{ij}^2} dA_j \tag{4.12}$$

The definition is now extended to whole surfaces:

$$F_{ij} = \frac{\varsigma_{ij}}{A_i} \iint \frac{\cos\theta_i \cos\theta_j}{\pi R_{ij}^2} dA_i dA_j \tag{4.13}$$

and the reciprocity rule is

$$A_i F_{ij} = A_j F_{ji} \tag{4.14}$$

Here ς_{ij} is 1 if surfaces i and j are visible to each other and 0 if they are not. As an example consider two surfaces labelled '1' and '2'. The energy leaving surface 1

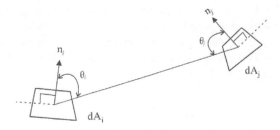

FIGURE 4.9 Two differential elements used for definition of view factor.

and arriving at surface 2 is $E_{b1}A_1F_{12}$, and the energy leaving surface 2 and arriving at surface 1 is $E_{b2}A_2F_{21}$. One of the simplest examples involves two infinite parallel plates, for which $F_{12} = F_{21} = 1$.

The summation rule, which applies to every surface is a statement of energy conservation, is

$$\sum_{j=1}^{N} F_{ij} = 1 \qquad (4.15)$$

where N is the number of surfaces. Extensive catalogues are available, in both graphical and analytical forms, e.g. (Howell and Menguc, 2011; Howell et al., 2021).

Means of calculating exchange between surfaces are of course discussed extensively in the cited literature, and outlined briefly in Section 4.1.5 and Appendix B, which summarises how the shape factors are employed.

4.1.5   INTRODUCTION TO THE ELECTRICAL ANALOGY

The so-called 'network method', also known as the radiosity or electrical analogy method, was originally proposed by Oppenheim (Oppenheim, 1956) for enclosures containing a transparent medium, and has been extended and generalised as discussed in the cited references (see also (Clark and Korybalski, 1974; Tao and Sparrow, 1985)).

The technique uses an analogy between radiative transfer and electrical networks or circuits containing resistances. This section only introduces a few basic concepts, and the technique is discussed further in Section 5.4.1 and Appendix B. The methodology is described neatly by (Holman, 2018), among others. The initial focus here is on transparent domains bounded by black or grey surfaces which are discretised to justify an assumption of constant temperature or radiative flux. Consider a grey surface at the boundary of a domain. The *radiosity* (denoted by J, in W/m^2) is defined as the energy per unit area leaving the surface into the domain (see Section 4.1.2 and Appendix B), and is thus given by

$$J = \epsilon E_b + \rho G_w \qquad (4.16)$$

Here G_w is the *irradiation*, i.e. the radiation arriving at the surface from the interior (per unit area), also known as the 'incident radiation' (but see Section 4.1.11). Now the net energy flow per unit area from the surface towards the domain, or radiative flux at the boundary, is the radiosity minus the irradiation. Noting Equation (4.8), the net radiative energy flow (in W) from the surface into the domain (for zero transmission) is thus given in terms of the radiosity by

$$Q = \frac{E_b - J}{(1-\epsilon)/\epsilon A} \qquad (4.17)$$

where A is the surface area. This is analogous to electrical current with the numerator being the potential (Voltage) and the denominator being the resistance. The network method leads to a system of linear, simultaneous, algebraic equations for the radiosity at each surface and from that deduces the radiative flux at that surface using Equation (4.17) (see Appendix B).

An alternative technique is the 'net radiation method', which leads to a system of linear algebraic equations for the net radiative flux at each surface instead of the radiosity. The classical method, for grey opaque enclosures, is described in the literature, e.g. (Sparrow and Cess, 1978; Rohsenow *et al.*, 1998; Lienhard and Lienhard, 2012; Modest, 2021; Howell *et al.*, 2021), as is the extension to enclosures which included both opaque walls and openings/windows, as devised by (Siegel, 1973). Both methods can handle enclosures which are entirely opaque, or which include some openings/windows. Different preferences are expressed in the literature, but given the omnipresence of computers and the availability of a range of radiation solvers in most CFD codes, there is little to choose between these approaches and this book focuses on the network approach within its limited non-CFD scope.

Both methods have been extended to simple categories of participating media, but this becomes tedious, especially in view of the availability of modern radiation solvers. Interestingly, though, the network method does manifest in some of the hybrid or combined radiation solver methods for non-transparent media (see Section 5.4.2.9). The network method with a single participating zone is described in (Holman, 2018), and generalisations are to be found in (Tong and Tien, 1980; Yuen, 1990), among others. The work of (Sinai *et al.*, 1993) is an example involving a single homogeneous zone (filled by a participating material) and multiple bounding surfaces, handled computationally. The spectroscopic properties of the medium in that case are determined by the characteristics of a suspended aerosol (calculated as part of the solution), and are computed parametrically in advance and stored in look-up tables, together with the surface-to-surface and surface-to-medium view factors (computed parametrically by the Monte Carlo method), thereby reducing the computing costs by a large factor. Modelling of the single-zone case with the net radiation method is described by (Stasick, 1988; Howell *et al.*, 2021).

4.1.6 RADIATION INTENSITY

A fundamentally important quantity is the spectral radiation intensity I_λ (also known as radiance in some sectors), introduced briefly in Section 4.1.2 (Equation (4.5)) and which is defined as the radiative energy flow in the given direction at a given point within the 3-D domain, per unit solid angle, per unit area normal to the ray, per unit time, per unit wavelength. The sum of I_λ over all wavelengths, known as the total radiation intensity, is thus the radiative energy flow per unit solid angle, per unit area normal to the ray, per unit time, with common units of $W\ m^{-2}\ sr^{-1}$, where sr stands for steradians. Here λ denotes wavelength in metres, although in heat transfer it is often quoted in microns (μm).

A schematic of the definition is provided in Figure 4.10. If the point P is at a boundary, then since the elemental area dA_1 is perpendicular to the intensity vector,

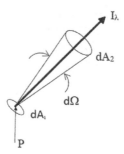

FIGURE 4.10   Schematic of intensity, shown by the bold arrow at a point P which may be on a boundary or within the domain. Similarly, the elemental area dA_1 may lie at or adjacent to a boundary, or within the domain. $d\Omega$ is the solid angle (steradians) subtended at P by the elemental area dA_2.

it will be close to the boundary but will only be part of the boundary for the direction perpendicular to the surface.

At any point inside the computational domain, the intensity can and usually does vary with direction. This is discussed further in Section 4.1.7 below.

4.1.7   RADIATIVE FLUX

Here the discussion deals with the radiative flux not at boundaries, but at points inside the computational domain, even if the concepts can be applied to boundaries too. Since the intensity is a function of direction, the net (resultant) spectral radiative flux locally at a point is a vector evaluated by summing vectors over all directions.

$$\vec{q}_\lambda = \int_{4\pi} I_\lambda\left(\vec{s}'\right)\vec{s}'\,d\Omega' \tag{4.18}$$

Here the dash denotes the dummy variable of integration, $\vec{s}'$ is the unit vector in the dummy direction during integration, and Ω' is the dummy solid angle. The component of the net spectral radiative flux in any (unit) direction $\vec{s}$, is a scalar (which can be positive or negative) denoted by $q_{\lambda s}$ and is given by

$$q_{\lambda s} = q_\lambda\left(\vec{s}\right) = \int_{4\pi} I_\lambda\left(\vec{s}'\right)\vec{s}\cdot\vec{s}'\,d\Omega' \tag{4.19}$$

In terms of rectangular coordinates, writing the unit vectors in the usual form as

$$\vec{s} = \mu_x i + \mu_y j + \mu_z k \tag{4.20}$$

where μ_x, μ_y, μ_z are the three direction cosines, with similar expressions involving dashes for $\vec{s}'$. Using spherical coordinates (see Appendix D), the three direction cosines are given by

$$\vec{s} = \begin{bmatrix} \sin\theta\cos\phi \\ \sin\theta\sin\phi \\ \cos\theta \end{bmatrix} \tag{4.21}$$

with a similar expression applied to the 'dashed' vector. The net flux in any direction $\vec{s}$ is thus

$$q_{\lambda s} = \mu_x q_{\lambda x} + \mu_y q_{\lambda y} + \mu_z q_{\lambda z} \tag{4.22}$$

Noting Equation (D.1) in Appendix D, the following apply:

$$q_{\lambda x} = \int_{\phi=0}^{2\pi} \int_{-\pi/2}^{\pi/2} I_\lambda\left(\theta',\phi'\right)\sin^2\theta'\cos\phi'\,d\theta'\,d\phi' \tag{4.23}$$

$$q_{\lambda y} = \int_{\phi=0}^{2\pi} \int_{-\pi/2}^{\pi/2} I_\lambda\left(\theta',\phi'\right)\sin^2\theta'\sin\phi'\,d\theta'\,d\phi' \tag{4.24}$$

$$q_{\lambda z} = \int_{\phi=0}^{2\pi} \int_{-\pi/2}^{\pi/2} I_\lambda\left(\theta',\varphi'\right)\sin\theta'\cos\theta'\,d\theta'\,d\phi' \tag{4.25}$$

The total flux is course derived by integrating over all wavelengths, and Equation (4.19) applies as it is for a grey situation. This relates to points inside the domain, and at boundaries the dot product of this vector is taken with respect to the surface normal and integrations are performed over hemispheres (exceptions occur at locations which are not smooth). As an example, for the case of a plane black surface lying in the x-y plane, these equations, together with Equation (4.5), yield the classic Lambertian or cosine law (which is deduced simply from the projected elemental surface area):

$$q_s = E_b \cos\theta \tag{4.26}$$

These expressions also relate to the divergence of the radiative flux, which quantifies the volumetric coupling between the radiation and thermal fields.

4.1.8 Diffusion, Anisotropy and Collimation

In general, the intensity at any point inside the domain will be a function of direction. Such a field is described as 'diffuse'. Note that in the context of solar radiation, 'diffuse' specifically refers to the shortwave radiation scattered by the atmosphere. That is also referred to as 'sky' radiation, and is supplemented by radiation emitted by the atmosphere, and radiation reflected and emitted by the ground.

In one extreme limit the intensity is constant across all directions, and is described as 'isotropic'. This situation arises in optically thick media well away from the boundaries. If the intensity does vary with direction the field is sometimes described as anisotropic or non-isotropic.

The other extreme situation is one in which the intensity is confined to, or close to, one direction, as occurs in solar radiation reaching Earth, searchlight or torchlight beams, and lasers. Such fields are described as 'collimated', or 'Direct' radiation in the context of the atmosphere and solar effects. Beams operating in the visible range are made visible due to scattering by particulates such as dust and/or water droplets, although molecular (Rayleigh) scattering can render the beam visible when air is nominally clear.

Unless the medium scatters anisotropically (cf. Section 4.1.10), or collimated radiation is involved, it is reasonable to assume that often (but definitely not always) anisotropy is dominated by wall effects, and that the participating medium tends to diminish that behaviour. An illustration of surface-driven anisotropy is provided in Figures 4.11 and 4.12, for the classical case of a transparent medium between two infinite, parallel black plates, labelled '1' and '2' respectively. Such a configuration is an approximation for a plane slab which is finite but has a high aspect ratio. In a typical spherical co-ordinate system the z axis is perpendicular to the slab, and the

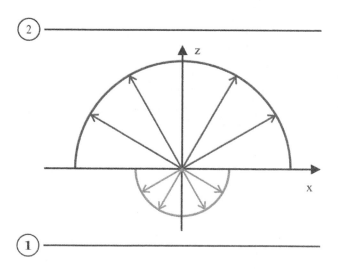

FIGURE 4.11 Polar diagram of intensity at any point inside an infinite, plane-parallel, transparent slab between two infinite, black plates labelled '1' and '2'. The origin of the polar plots is at the intersection of the two axes, which is at the point being considered. The units are W m^{-2} sr^{-1}. $T_1 = 100°C$ and $T_2 = 30°C$. The top arc denotes radiation with directions having a positive component in the z direction, and the lower arc applies to directions with a negative component. The radius of the top hemisphere is 349.9 W m^{-2} sr^{-1}, and that of the lower hemisphere is 152.4 W m^{-2} sr^{-1}.

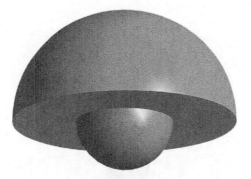

FIGURE 4.12 3-D view of the angular intensity distribution at a point inside the transparent slab between black walls, with the lower wall at 100°C and the top one at 30°C.

zenith θ is measured from the z axis. Thus radiation travelling from 1 to 2 has zenith lying in the range $0 \leq \theta < \pi/2$, shown in red, and radiation travelling from 2 to 1 has zenith in the range $\pi/2 < \theta \leq \pi$, shown in blue. The example shown here applies to one diffuse plate at 100°C and the other at 30°C. Thus, the 3-D polar surface consists of two hemispheres (see Equation (4.5)), the top one with radius = 349.9 W m^{-2} sr^{-1}, and the lower having a radius of 152.4 W m^{-2} sr^{-1}. This is consistent with the radiative transfer equation (Section 4.2), which states that the intensity along a ray is constant in a transparent medium.

Figure 4.13 shows the polar distribution of the flux for this eample, which from Equations (4.23) to (4.25), is axisymmetric about the z axis, meaning that only the z component of q_s is non-zero. For this example, for any direction (ϕ, θ) the horizontal component of the flux is exactly balanced by the component of the flux at $\phi + \pi$ in the same hemisphere. The distribution is Lambertian, which in 2-D is a circle of diameter 1099.4 W m^{-2} for directions in the positive z direction (centred at 549.7 above the x axis), and 478.8 W m^{-2} for the negative z directions (centred at 239.4 below the x axis).

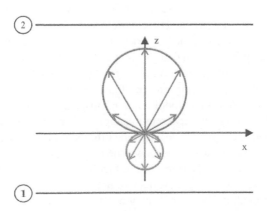

FIGURE 4.13 Polar diagram of the radiative flux at any point inside the transparent slab.

FIGURE 4.14   3-D view of the angular distribution of the radiative flux at any point inside the transparent slab.

If the slab is not transparent then the flux distribution is affected of course, and along the positive or negative x directions the flux is non-zero.

4.1.9 INTERFACES AND REFRACTION

In Section 4.1 above, discussion dealt with radiative properties at a surface separating one medium from another. In the typical process engineering context, the hatched regions below the surface in Figure 4.5 represent an opaque medium such as metal. Indeed, by opaque we mean that the radiation is absorbed over very small distances. In such cases, in addition to some of the radiation arriving at the interface from the non-opaque side being reflected, the remainder is absorbed in its entirety close to the surface of that material and heat is then transported in that material by other mechanisms. A different situation is one in which neither material is opaque. If the refractive indices (see Section 4.1.3) of the two materials separated by the interface are not equal, refraction will take place, whereby the direction of propagation of the electromagnetic waves (i.e. the direction normal to the wave fronts) changes as the waves pass through the interface. The classic Snell's law states that when passing into a medium with a higher n, the wave direction moves closer to the interface normal, and vice versa. Whether that energy escapes from the second material elsewhere depends on the specifics of its radiative properties and geometry. Glass is an example of semi-transparent materials, non-grey in this case. The reader is referred to physics, optics and some radiation textbooks, e.g. (Rohsenow *et al.*, 1998; Modest, 2021; Howell *et al.*, 2021) for the classical relationships such as Snell's law, Fresnel's equation, the Critical and Brewster angles, and the effects of oblique wave incidence and of polarisation, but two of these aspects will be cited in this book.

Consider a plane radiatively smooth interface separating two media, denoted by '1' and '2' and possessing complex refractive indices m_1 and m_2. A wave approaches the interface from medium '1', in a direction which is at an angle θ_1 to the interface normal (see Figure 4.15).

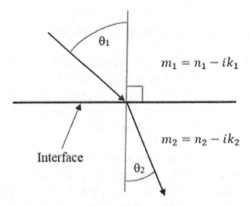

FIGURE 4.15   Refraction at a plane interface separating two media.

The wave is refracted at the interface due to the change in wave speed, and because the wave phases in the two media coincide at the interface, so that its direction is altered. The classic Snell's law states that

$$\frac{\sin\theta_2}{\sin\theta_1} = \frac{n_1}{n_2} \tag{4.27}$$

First, consider reflection and transmission when a wave approaches a plane, optically smooth interface, separating non-absorbing dielectrics (poor conductors), with the wave incidence normal to the interface. The regions straddling the two sides of the interface possess refractive indices n_1 and n_2, with $n_2 > n_1$.

$$\hat{R}_n = \left(\frac{n_2 - n_1}{n_2 + n_1}\right)^2; \hat{T}_n = \frac{4n_1 n_2}{\left(n_2 + n_1\right)^2} \tag{4.28}$$

Here $\hat{R}_n$ and $\hat{T}_n$ are the ratios of the reflected and transmitted powers to the power of the (normal) incident field, and by definition their sum equals 1:

$$\hat{R}_n + \hat{T}_n = 1 \tag{4.29}$$

For a typical air-glass interface $n_1 \approx 1$ and $n_2 \approx 1.5$, whence

$$\hat{T}_n \approx 96\%; \hat{R}_n \approx 4\% \tag{4.30}$$

This also applies to the wave travelling in the opposite direction, from glass to air. The behaviour at oblique angles (i.e. angle between the ray direction and the normal is non-zero) is more complex, and some information is provided in Appendix C

('Fresnel's equations'), but the reader seeking detailed information should consult the literature already cited. The reflectivity in the above case increases rapidly as the incidence angle increases towards 90°.

Behaviour within a slab is also more complicated of course and involves multiple reflections at the two interfaces. The literature includes work on the effect of the slab's refractive index, e.g. (Rokhsaz and Dougherty, 1989; Siegel and Spuckler, 1992; Howell *et al.*, 2021). Nonetheless, it is worth quoting some results for a three-layered non-absorbing system, in which there are two interfaces, essentially involving a slab separating two semi-infinite regions (Howell *et al.*, 2021). This is relevant to thin coatings on glass, so the situation addressed here can involve air adjacent to a thin coating which sits on a thick substrate (e.g. glass). Denote the three refractive indices by n_1, n_2, n_3. The normal reflectivity when the film/coating is much thicker than the wavelength, and wave interference can be ignored, is

$$\hat{R}_n = 1 - \frac{4n_1 n_2 n_3}{\left(n_2^2 + n_1 n_3\right)\left(n_1 + n_3\right)} \tag{4.31}$$

This is minimised (and transmission maximised) when

$$n_2 = \sqrt{n_1 n_3} \tag{4.32}$$

For air and glass, the best that can be achieved is reduction of reflection to about a half of the uncoated value. However, much lower reflection can be achieved if the coating thickness is comparable or smaller than the wavelength, when interference cannot be ignored. Indeed, it turns out that if the coating thickness is a quarter of the wavelength in the coating, then the normal (and monochromatic) non-absorbing reflectivity is zero if the coating refractive index is equal to the geometric mean in Equation (4.32). Conversely, large reflectivity is achieved when n_2 is large.

4.1.10 SCATTERING

Scattering is the redirection of energy due to inhomogeneities in the medium. This can be associated with very small scales, involving molecules, or larger scales involving different materials, such as dispersed particles carried in a fluid (e.g. water droplets, dust, or soot) or heterogeneous materials making up a porous medium (see Sections 5.3.6 and 5.3.7).

Scattering by a single, homogeneous non-spherical particle is illustrated in Figure 4.16. Three aspects play a role:

   a. Reflection at the interface between the particle and the surrounding medium.
   b. Refraction as the photons pass through the interface.
   c. Diffraction, whereby the photons are deflected without contacting the obstacle.

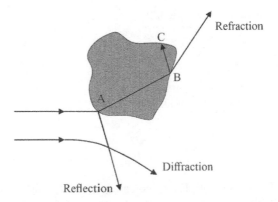

FIGURE 4.16 Sketch of scattering by a particle (shaded). Photons arriving from the left. Refraction and reflection occur at A and B. The photons reflected at B continue towards C, where reflection and refraction occur too, and the process continues repeatedly thereafter, and is not shown here. Adapted from (Modest, 2021).

Absorption may also take place inside the particle, although that does not alter the photon redirection. At point A, the first reflection and refraction occurs for the ray shown approaching from the left. At point B, photons are refracted to leave the particle, and reflected to arrive at point C, where reflection and refraction occur again, and the process is then repeated ad infinitum. Thus, photons are potentially scattered in all directions: Forward, laterally, and even backward. The frequency of a small proportion of cases is altered, an effect described as 'inelastic' scattering, but this aspect can be ignored in heat transfer applications, which are the focus of this book.

If additional particles exist, they may interfere with these phenomena. If the separation between the particles, defined below, is sufficiently small, such interference exists and the scattering is said to be *dependent*. Conversely, if the separation is sufficiently large, the interference is negligible, and the scattering is said to be *independent*.

In view of all the above, the literature indicates that the scattering at a cloud of particles depends on the geometry of particles or inhomogeneities, and two parameters. For a cloud made up of spherical particles, this leads to four dimensionless parameters as follows. 'Separation' refers to the size of the nearest gap between the particles.

Complex index of refraction (Equation (4.10)):

$$m = n - ik \tag{4.33}$$

Ratio of particle circumference to wavelength:

$$\hat{d} = \frac{\pi d}{\lambda} \tag{4.34}$$

Ratio of particle separation to wavelength:

$$\hat{L}_p = \frac{L_p}{\lambda}$$ (4.35)

Ratio of particle separation (i.e. gap) to particle diameter:

$$\tilde{L}_p = \frac{L_p}{d}$$ (4.36)

The review by (Tien and Drolen, 1987) indicates that scattering is independent when $\hat{L}_p > 0.5$ and $f_v < 0.006$, where f_v is the particle volume fraction, and then only the first two of these parameters apply (Equations (4.33) and (4.34)).

Focusing on independent scattering first, a general theory of scattering of plane waves by a single, homogeneous, spherical particle covering the complete range of d exists, and has traditionally been named after Mie, but in recognition of others who developed the subject independently, is nowadays named as Lorenz–Mie or Lorenz–Mie–Debye e.g. (van de Hulst, 1981; Grandy Jr and Grandy, 2005; Bohren and Huffman, 2008). However, approximations exist for small and large d, involving much simpler mathematics, and three regimes are usually identified:

i. $\hat{d} \ll 1$, Rayleigh scattering. Scattering is inversely proportional to the fourth power of wavelength, which led to the classical explanation of blue sky (at shorter wavelengths) and red sunsets (at longer wavelengths).
ii. $\hat{d} = O(1)$, Lorenz–Mie–Debye scattering.
iii. $\hat{d} \gg 1$, geometric optics. The behaviour at the interface is well approximated by behaviour at a plane interface.

The theory for non-spherical particles is more complex. Turning now to dependent scattering, for a system possessing a general topology, the theory is even more complicated, and the reader is referred to Sections 5.3.6 and 5.3.7 below for introductory information on particulates and porous media in a variety of applications.

These theories can be used to derive the scattering phase function deployed in the radiative transfer equation (see Section 4.2); they will not be quoted here and may be found in the literature already cited. More is said about scattering in Section 4.2 and Appendix D.

4.1.11 THE MEANING OF 'INCIDENT' AND 'MEAN' RADIATION OR INTENSITY

Most undergraduate courses in engineering only touch briefly on the effects of participating media, and the term 'incident radiation' is usually only encountered in the context of surface-to-surface radiation. It is worth discussing the meaning of this term in the more general context.

When the problem only involves transparent materials inside the domain, with only surface-to-surface radiation, the term 'incident radiation' usually refers to the radiation (W m^{-2}) arriving at a surface (which is a boundary) from other parts of the boundary. This is also known as the 'irradiation' – see Figure 4.8 – and is denoted by G_w in Section 4.1.5. For a point on a smooth surface, this radiation arrives from a hemisphere (2π steradians).

If the problem involves boundary surfaces as well as a participating medium, then in addition to the surface irradiation, the term incident radiation is also used in relation to points inside the computational domain, away from the boundaries. At such locations the radiation arrives from all directions, i.e. from 4π steradians. It is common for CFD codes to deliver predicted fields of the total incident radiation at all points, which is the integral of the intensity (at each point) over all directions, and perhaps the directionally averaged incident radiation (which is the total divided by 4π). The term 'total' here means the intensity integrated over all directions at the given point, although it should be remembered that that term may also refer to the sum over all wavelengths or frequencies.

The spectral incident radiation at points inside the domain, irrespective of the opacity, is denoted here by G_λ and is given by

$$G_\lambda = \int_{4\pi} I_\lambda d\Omega \tag{4.37}$$

Here $d\Omega$ is the elemental direction. The spectral mean intensity at points inside the domain is thus

$$\bar{I}_\lambda = \frac{G_\lambda}{4\pi} = \frac{1}{4\pi} \int_{4\pi} I_\lambda d\Omega \tag{4.38}$$

The term 'mean intensity' is usually understood to involve the total value in the spectral sense, i.e. the sum of the spectral mean intensity over all wavelengths, at the given point:

$$\bar{I} = \frac{1}{4\pi} \int_{0}^{\infty} \int_{4\pi} I_\lambda \, d\Omega \, d\lambda \tag{4.39}$$

The incident radiation at points inside the computational domain is important because it forms part of the coupling between radiation and flow, as discussed in Section 4.4 below. It can also be helpful during post-processing. For example, the mean intensity is also used in some measures of thermal comfort (see Section 4.1.12), and it provides a rough estimate of the radiative flux delivered to a sub-grid object, at locations where no explicit geometrical representation exists (see Section 5.5).

4.1.12 MEAN RADIANT TEMPERATURE

Following on from the earlier discussion of total and mean intensities, an entity relevant to the assessment of thermal comfort is addressed briefly. The principal application of thermal comfort is in the built environment (and other situations too), and affects energy consumption and emissions. The abovementioned entity is named the Mean Radiant Temperature (MRT), and is the temperature of a fictitious homogeneous black surface surrounding the point being considered which would produce the same total radiant energy delivered to the point as is delivered in the actual, real situation. It is important in buildings when designing for energy efficiency and indoor air quality, but is also helpful in understanding the radiative environment inside the domain being analysed in other spheres of activity. By definition, in terms of the mean radiation intensity at a point inside the computational domain, it is formally given by

$$T_R = \left(\frac{\pi \bar{I}}{\sigma} \right)^{1/4} \tag{4.40}$$

The overview by (Sinai *et al.*, 2016) discusses this and cites approximations used in the built environment sector and which tend to be less accurate than this definition.

4.2   INTRODUCTION TO THE RADIATIVE TRANSFER EQUATION AND OPACITY

The radiation field, when heat transfer is being analysed, is governed by a single scalar equation known as the Radiative Transport Equation (RTE). Besides the usual derivation, some of the literature which has already been cited includes derivations of the RTE from the fundamental Maxwell equations. In mathematical terms, the RTE is an integro-differential equation for the radiation intensity I, which is a function of wavelength, direction, spatial position, and time, as discussed above. In the absence of a transient term which pertains in rapid phenomena such as pulsed lasers or relativistic phenomena, and is usually negligible in radiative heat transfer, the RTE is

$$\vec{\Omega} \cdot \nabla I_\lambda + K_\lambda I_\lambda = K_{a\lambda} n^2 I_{B\lambda} + \frac{K_{s\lambda}}{4\pi} \int_{4\pi} I_\lambda \left(\vec{x}, \vec{\Omega}' \right) \tilde{p}_\lambda \left(\vec{\Omega}, \vec{\Omega}' \right) d\vec{\Omega}' \tag{4.41}$$

where I_λ is the spectral radiation intensity, $I_{B\lambda}$ is the spectral blackbody intensity (see Figure 4.2 and Equation (4.5)), and $\vec{\Omega}$ is direction. In heat transfer applications, radiation can be regarded as unpolarised and incoherent. The RTE involves important spectroscopic/optical properties of the medium: K_λ is the *extinction coefficient* (m^{-1}), describing the change along a ray due to absorption and scattering, given by

$$K_\lambda = K_{a\lambda} + K_{s\lambda} \tag{4.42}$$

where $K_{a\lambda}$ and $K_{s\lambda}$ are the *absorption and scattering coefficients* respectively, which generally depend on the wavelength of the radiation field. These two coefficients are also described as the *absorption and scattering cross section* (m^2) per unit volume respectively. In the molecular context, the cross-sections are defined as the probability that a photon will be absorbed or scattered, multiplied by the projected cross-sectional area of the molecule, with an analogous definition in the context of particulates. The *absorption and scattering efficiencies* or *efficiency factors* (which are dimensionless) are defined as the relevant cross section divided by projected area of the molecule or particle, which is equal to the relevant probability. The *mass-specific absorption, scattering and extinction coefficients* are the relevant cross-sections per unit mass (usually quoted in m^2/g), so that the absorption, scattering and extinction coefficients are the relevant mass-specific coefficients multiplied by the mass concentration or density. The single scattering albedo is defined as the ratio of the local scattering to extinction coefficients:

$$\omega_\lambda = \frac{K_{s\lambda}}{K_\lambda} \tag{4.43}$$

The entity $\tilde{p}_\lambda$ is known as the *scattering phase function*, which depends on the scattering angle, in general also on direction and position, and is a spectroscopic property of the medium. It represents the probability that a photon arriving along direction $\vec{\Omega}'$ will be scattered in the direction $\vec{\Omega}$ being analysed by the RTE, Equation (4.41). The term *conservative scattering* relates to pure scattering, with the albedo = 1.

In descriptive form, the RTE (Equation (4.41)) can be stated as follows:

Rate of change of the spectral intensity in the given direction (with a positive sign meaning increase in that direction, and a negative sign meaning the opposite) =

+ *emission (1st term on right-hand side (RHS))*
− *attenuation due to absorption (absorption component of 2nd term on left-hand side (LHS))*
− *attenuation due to scattering out of the given direction (scattering component of 2nd term on LHS, known as out-scattering).*
+ *augmentation by scattering into the given direction of radiation arriving from all directions (2nd term on RHS, known as in-scattering).*

It is the scattering which introduces the greatest complexity of the integral element of the integro-differential character of the RTE, embedded in the last term in Equation (4.41). Note, however, that the complication of directionality exists even in the absence of scattering, due to the need to calculate directional integrals for the coupling with flow, and at boundaries. It is common in heat transfer to assume that $\tilde{p}$ is only a function of the angle between $\vec{\Omega}'$ and $\vec{\Omega}$. Exceptions may, for example, involve structures which are anisotropic due to their morphology, like fibres and crystals.

Generally, the phase function is expanded as a truncated series of Legendre polynomials P_j (Abramowitz and Stegun, 1965). The reader is referred to the cited textbooks for details, and discussion here will be limited. Referring to Appendix D, the expansion with N terms is

$$\tilde{p}_\lambda\left(\mu_\psi\right) = \sum_{j=0}^{N} a_j P_j\left(\mu_\psi\right) \tag{4.44}$$

The simplest case is isotropic scattering, with $a_0 = 1$ and $N = 0$. The next level is $N = 1$, called *linearly anisotropic*, i.e.

$$\tilde{p}_\lambda \left(\mu_\psi \right) = 1 + a_1 \mu_\psi \tag{4.45}$$

Many CFD codes allow for isotropic scattering, and some for linearly anisotropic scattering, or more refined models. A little more information is provided in Appendix D.

A useful measure of anisotropy is the *mean cosine of scattering*, or *asymmetry parameter*, defined as

$$\bar{\mu} = \frac{\displaystyle\int_{4\pi} \cos\psi\, \tilde{p}\left(\cos\psi\right) d\Omega}{\displaystyle\int_{4\pi} \tilde{p}\left(\cos\psi\right) d\Omega} \tag{4.46}$$

This varies between -1 and $+1$, with zero corresponding to isotropy. Positive values correspond to larger forward scatter, and the closer it is to 1.0 the larger the forward peak is compared with the backward scatter peak. Negative values correspond to backscatter, and the closer it is to -1 the larger the backscatter peak.

The refractive index also appears in Equation (4.41). Also, there may be other radiative sources, for example from some chemical or electrical processes. That will be ignored in this book.

The RTE is of course subject to boundary conditions, and the mathematical details depend on the software capabilities and can be found in the cited textbooks and in software user manuals. For a given direction, the RTE is a first-order differential equation, and it is logical to specify the single value that it requires as a starting value at a surface bounding the medium. In other words, the radiosity is specified at boundaries. We begin with a descriptive form which encompasses most situations, including opaque and transparent or semi-transparent boundaries. The code computes the intensity arriving at the boundary from the interior, which can be a function of location, direction, and wavelength. The boundary condition sets the radiosity at the same location. In descriptive terms,

Radiosity = surface diffuse emission + diffuse reflection + specular reflection + diffuse external source + collimated external source

The external source modelling depends on the functionality available in the CFD code, and will be ignored in this section. First, consider a surface which is a diffuse reflector. Mathematically, the above is then stated as follows (note the relations in Appendix D for directional integrations):

$$I_\lambda \left(\vec{s} \right) = \varepsilon_\lambda \left(\vec{s} \right) I_{b\lambda} + \frac{\rho_\lambda}{\pi} \int_{2\pi} I\left(\vec{s}' \right) \left| \vec{n} \cdot \vec{s}' \right| d\Omega' \tag{4.47}$$

Here ρ is the reflectivity, $\vec{n}$ is the local surface normal pointing into the interior, and the integration is performed over the hemisphere on the interior side of the boundary, i.e. $\vec{n} \cdot \vec{s}' < 0$. Thus, the second term is the summation over all directions of the irradiation, arriving from the interior, multiplied by the local reflectivity. If the surface is a diffuse emitter, then its emissivity in the first term is constant at the given wavelength.

If the reflectivity is hybrid, as depicted for example in Figure 4.6, then the reflectivity in the integral in Equation (4.47) is the diffusive component, written as $\rho_{\lambda d}$, and a third term is added to the right-hand side:

$$\text{Specular term} = \rho_{\lambda s} I \left[\vec{s} - 2 \left(\vec{s} \cdot \vec{n} \right) \vec{n} \right] \tag{4.48}$$

where $\rho_{\lambda s}$ is the specular reflectivity. Generalisations are discussed in (Modest, 2021; Howell *et al.*, 2021).

Additional aspects of boundary conditions are discussed in Section 5.1.

The (spectral) optical thickness or depth for absorption, when the properties are inhomogeneous, is given by

$$\tau_\lambda = \int_0^s K_{a\lambda} \left(s' \right) ds' \tag{4.49}$$

where s is distance along a geometric path. Analogous expressions apply to the optical depths for scattering and extinction. The extinction optical depth is sometimes described in terms of 'opacity', although that term is often associated with a typical or characteristic extinction coefficient K as follows:

$$\tau = KL \tag{4.50}$$

Here L is a characteristic spatial dimension of the geometry. Thus for zero or very small opacity the medium is transparent (or nearly so), if the opacity is much larger than 1 the medium is said to be optically thick or opaque (and participating), and if the opacity is $O(1)$ then the medium is participating.

Scattering increases the mathematical complexity of the RTE considerably. The scaling theories in (Lee and Buckius, 1982, 1983) are able to simplify some scattering cases:- They reduce an isotropically scattering problem (i.e. scattering is independent of the scattering angle) to a non-scattering one, and a linearly scattering one (i.e. scattering is a linear function of the scattering angle) to an isotropically scattering one.

Conversely, the RTE is simplified significantly if scattering is absent, and then becomes a differential equation in CFD applications, although even here the equation is an integro-differential one in the so-called state of 'radiative equilibrium' (see below and Section 4.5). Ignoring that particular situation, the mathematics is still complicated by the dependence of the intensity on direction. Clearly, the term 'scalar' applied to the character of the RTE belies its great complexity due to the dependence on direction.

If absorption and emission within the medium can also be ignored the situation is simplified further to surface-to-surface (transparent) scenarios. Few non-transparent cases can be solved analytically, and those may be found in the literature which has already been cited. A good review is given by the online appendix of (Howell *et al.*, 2021), where the analytical work of (Crosbie and Schrenker, 1982, 1984) is described as the best reference for checking methods in 2-D and 3-D respectively. Two more exact solutions (among others available in the literature) are given in Appendices E and F, for the infinite plane-parallel grey slab bounded by opaque, grey and diffuse boundaries with different emissivities and temperatures. The first case (Appendix E) is the homogeneous one in which the medium's temperature and properties are constant and specified, which implies multiple modes of heat transfer in which the radiative field is not in equilibrium. The second case (Appendix F) implies radiative equilibrium, with radiation overwhelming other modes of heat transfer, and the temperature predicted as part of the solution instead of being assumed.

Much of the early work emerged from astrophysics. In the engineering sphere significant effort was expended in deriving closed-form solutions for simple domain shapes, although there was a tendency to analyse equilibrium situations in which radiation is the only or the dominant mode of transfer, and volumetric emission is exactly balanced by volumetric absorption. In reality, problems involving participating media include coupled heat transfer, in which even an isothermal assumption is occasionally a reasonable approximation. Worthy of specific mention is the analytical work of (Williams, 1983, 1984), which yielded analytical results for the infinite plane slab including scattering. In the optically thick limit the equations of that theory degenerate to two coupled transcendental equations depending solely on the scattering albedo (K_s/K), cf. Appendix H. At the 2-D level, the work of (Duo and Azmy, 2007) is an example of a benchmark involving inhomogeneous spectroscopic properties.

There are several 'formal' (integrated) solutions to the RTE which are also worth mentioning, even if their applicability is limited; these are quoted in Appendix I, and will be cross-referenced in discussions below. Due to their importance to the immediate discussions, the reader is specifically referred to the Beer law (Equation (I.1)), which applies when only absorption takes place, to Equation (I.2) which applies when emission occurs too, and to Equation (I.3) which applies when isotropic scattering is present too.

Numerical approaches are common nowadays. These will be discussed below, as will the issue of the spectroscopic properties, embedded in the extinction coefficient and the refractive index.

4.3   ESTIMATION OF A CHARACTERISTIC OPACITY, AND DIMENSIONLESS GROUPS

The medium's opacity depends on its spectroscopic properties. If the latter can be reasonably approximated by an average value, then it is possible to estimate the 'characteristic' opacity based on the estimated average concentrations and other parameters which all affect the spectroscopic properties. This can provide a guideline as to whether the problem can or cannot be assumed to be a transparent one.

In the context of engineering combustion of hydrocarbons, common contributors to radiation are water vapour, carbon dioxide, and soot. Most heat transfer textbooks (including those already cited) provide graphs for the volume emissivities (ε_g) of various shapes of a homogeneous medium containing water vapour, or carbon dioxide, or both. An example of scoping estimates, for air containing water vapour and carbon dioxide under typical ambient conditions, is given in (Sinai *et al.*, 2016). The literature provides a characteristic radiative length, known as the Mean Beam Length (MBL) and denoted here by L_{mb}, for a range of generic shapes of the domain. In the typical application, the MBL is the length which, when multiplied by the extinction coefficient in the Beer law, yields the actual mean incident radiation at a boundary. A common approximation is (Rohsenow *et al.*, 1998)

$$L_{mb} = \frac{3.6\,V}{A} \qquad (4.51)$$

where V and A are the volume and total inward-facing area of the domain boundary, respectively. For air, when this is combined with the data on concentrations of water vapour and CO_2, Hottel diagrams provide the effective volume emissivity of the complete fluid domain, denoted here by ε_g. Analytical versions of the diagrams are available in the literature. The opacity τ_g is defined by

$$\tau_g = KL_{mb} \qquad (4.52)$$

where, as discussed earlier,

$$K = K_a + K_s \qquad (4.53)$$

Here K_a and K_s are the absorption and scattering coefficients, respectively. If the opacity is small, the medium is said to be optically thin (or transparent if relevant), if it is large the medium is said to be optically thick, and if the opacity is of order (1), some of the literature uses the term 'self-absorbing'.

From the computed volume emissivity, the opacity may be estimated to be

$$\tau_g = -\ln\left(1 - \epsilon_g\right) \qquad (4.54)$$

Note that for a given domain geometry, the volume emissivity will depend on the surface (making up part of the boundary) which is being considered. For the reader's benefit, the Antoine correlation for the saturated vapour pressure of water is supplied in Appendix F.

Taking this issue further, a few dimensionless groups are relevant (see for example (Kunes, 2012; Howell *et al.*, 2021)). The Boltzmann or Thring number is

$$Bo = N_{Th} = \frac{\rho C_P U}{\varepsilon \sigma T^3} \qquad (4.55)$$

where the characteristic parameters are ρ (density), C_P (specific heat capacity), U (advection speed), ε (emissivity of gas or boundary), and T (temperature). This number is the ratio of energy transfer by advection to energy transfer by radiation. Another dimensionless group specific to phenomena at boundaries is the Radiation Number, also called the Hottel Number:

$$N_R = \frac{h\left(T_g - T_w\right)}{\varepsilon\sigma\left(T_g^4 - T_w^4\right)} \tag{4.56}$$

where h is a convective heat transfer coefficient, T_g is a gas temperature, T_w is a boundary temperature, and ε is a combined wall-gas emissivity, equal to the wall emissivity if the gas is optically thick (see Appendix B). If the gas is not optically thick then engineering judgement is required, and one possible approximation is

$$\varepsilon = \left(\frac{1}{\varepsilon_w} + \frac{1}{\varepsilon_g} - 1\right)^{-1} \tag{4.57}$$

where ε_w and ε_g are the wall emissivity and an estimated gas emissivity respectively. N_R can be simplified to the following when the temperature differences are small:

$$N_R = \frac{h}{4\varepsilon\sigma T_m^3} \tag{4.58}$$

where T_m is the mean temperature $0.5(T_g + T_w)$.

Another obvious group is the opacity or optical depth (sometimes known as the Bouguer number):

$$\tau = KL \tag{4.59}$$

which has already been introduced in Section 4.2.

These dimensionless groups provide guidance as to whether surface-to-surface and volumetric radiation are significant.

4.4   COUPLING BETWEEN FLOW AND RADIATION

A key question for the flow analyst is whether to model radiation, and if so, how.

Much work has been done on 'combined' effects, involving both flow and radiation; for reviews, see (Ozisik, 1985; Modest, 2021; Howell *et al.*, 2021). A useful discussion of the issues is provided in (Tencer and Howell, 2016). Radiation can obviously affect the thermal field and the flow, and conversely, the flow, and especially the temperature, can affect the radiation. Generally, for non-transparent cases this coupling is more pronounced in natural convection than in forced convection, and the former has therefore received more attention. An early example of forced flow is (Chen, 1963). In recent years, the canonical forced convection problem of Poiseuille flow in a channel, and developing boundary layer along a flat plate, have received renewed attention, exploiting advances in computing power and turbulence modelling, and have identified the impact of radiation on the classical similitudes and log laws (Section 2.4). Zhang et al (2013) say that for the combined forced flow the usual temperature profile and its corresponding log-law are generally not valid within a turbulent boundary layer.

The magnitude of radiative heat transfer compared with other mechanisms has been introduced in Section 1 and Section 4.3.

The concept of radiatively participating media has also been introduced above, and demands a decision by the modeller. If the medium is non-transparent, then radiation affects the fluid flow, since the radiation manifests as a volumetric heat source (or sink) throughout the medium, and thus has the potential to affect the fluid temperature, and other entities such as density, and buoyancy forces if buoyancy plays a role. This is named here as 'volumetric coupling'.

Radiation can also affect flow by influencing boundary temperatures, since those boundary temperatures also affect the fluid density (and/or other properties of the fluid) close to the boundaries. These phenomena apply whether the medium is transparent or non-transparent. Thus, if the boundary temperatures are computed by the whole CFD model, as part of the solution process, then the flow is coupled to the radiation, and in general radiation should be computed (although as discussed, in some cases its effects can be weak). This aspect is named here as 'boundary coupling'.

On the other hand, if all the boundary temperatures are known, or prescribed a priori, then the radiative field is uncoupled if the medium is transparent; the flow can then be computed as convection only, and radiation assessed as a post-processing step, i.e. radiation can be solved on its own, given the thermal results of the CFD simulation.

Boundaries may of course be opaque or non-opaque.

Note that the boundary temperatures are computed, rather than prescribed, for any thermal boundary condition other than a prescribed temperature (a Dirichlet condition in mathematical terms); that includes adiabatic (insulated) boundaries, prescribed heat flux, or a prescribed heat transfer coefficient to a prescribed source/sink temperature. Sometimes, the temperature of the structure is predicted using a separate structural finite-element (FE) or zone model. If the FE/zone software does not itself account for radiation or does not do so realistically, then errors will arise; the modeller then needs to decide whether the errors are acceptable, and if not, whether different software or methodology needs to be deployed.

To summarise:

- Coupling between the thermal and radiative fields occurs through two mechanisms: Volumetric coupling and boundary coupling.
- If the medium is non-transparent, radiation is coupled to the flow and the two should be analysed simultaneously, irrespective of the type of boundary conditions. There may be flow regimes where this will not be essential. For example, if the Boltzmann or Thring numbers are very large (see Section 4.3), then advective heat transfer dominates radiation, and radiation

could be analysed separately, provided convection dominates radiation at the boundaries too, not only in the fluid bulk. The latter issue is guided by the Radiation Number (Section 4.3).

- If the medium is transparent and the boundary temperatures are specified a priori, then flow and radiation are uncoupled; the flow can then be analysed as a purely convective process, and radiation analysed separately.
- If the medium is transparent and the boundary temperatures are not specified in advance, and are rather computed as part of the solution, then generally radiation and flow are coupled, and should be analysed simultaneously.

It is worth noting that coupling the radiation and flow solvers does no harm – it is the most rigorous approach – but in some cases it can be superfluous and unnecessarily expensive.

Now consider volumetric coupling in more detail. The flow solution of course delivers parameters such a temperature, species concentrations, and soot concentrations, and these control the spectroscopic properties as discussed below, and thereby the radiative field. The second volumetric coupling mechanism exists as a source term in the energy equation, which is the radiative energy absorbed by the medium at the given point, minus the energy emitted at that point. Consider, for example, the enthalpy equation for flow in the context of eddy-viscosity turbulence modelling, Equation (2.17). The source terms on the right-hand side include a radiative contribution equal to $-\nabla \cdot q_R$, where q_R is the local, total radiative flux, i.e. the radiative flux summed over all wavelengths. For non-grey material, the radiative source is

$$S_R = -\nabla \cdot q_R = -\int_0^\infty K_{a\lambda}\left(4\pi I_{B\lambda} - G_\lambda\right)d\lambda \qquad (4.60)$$

Here $K_{a\lambda}$ is the spectral absorption coefficient, and G_λ is the spectral incident radiation at the given point, i.e. the integral of the intensity at the given point over all directions:

$$G_\lambda = \int_{4\pi} I_\lambda\left(\vec{x},\vec{\Omega}'\right)d\vec{\Omega}' \qquad (4.61)$$

Thus, the enthalpy source term is equal to the radiative absorption minus the radiative emission; the former is lost from the radiative field and delivered to the convective field, and the latter does the converse. For a grey gas, this becomes

$$-\nabla \cdot q_R = -K_a\left(4\sigma T^4 - G\right); \; G = \int_0^\infty G_\lambda d\lambda \qquad (4.62)$$

As already explained, this is a negative term representing energy loss by radiative emission (the first term on the right-hand side), and a positive source representing energy gain by absorption of radiation (the second term on the RHS). These

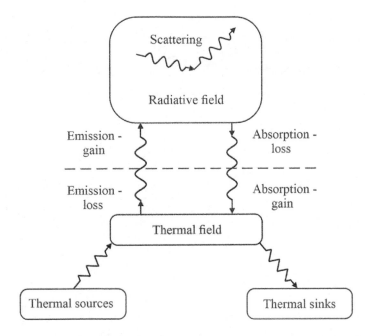

FIGURE 4.17 Schematic of the local volumetric coupling between flow (via the thermal field) and radiation, for a single-phase medium. A wave symbol represents radiative sources, and a resistor represents convective sources.

processes are shown in Figure 4.17, depicting the local volumetric coupling for a single-phase medium. Scattering does not affect the coupling directly, as discussed. If additional radiative internal sources exist they will be connected to the top box, but are ignored in this book. The behaviour is illustrated nicely by an example of a well-mixed domain in which the temperature is essentially uniform apart from thin boundary layers at boundaries. That temperature can, for example, be affected by the boundary temperatures and flow inlets and outlets. A distinction has been made here between 'sources' and 'sinks', but these can equally be described as 'sources' possessing positive or negative values, respectively.

The important and obvious fact to note is that generally the thermal field is affected by both radiation and by non-radiative heat transfer. This will be elaborated on in the section on radiative equilibrium.

In general boundary coupling exists too of course, and Figure 4.18 is the same as the above but with the addition of exchanges between the local flow element and elements of the boundary. Non-local effects also include exchange between boundary elements, which is influenced by the contained medium and which degenerates to the classical surface-to-surface exchange when the medium is transparent.

It is also important to consider a situation which is not uncommon, namely multi-phase flow (cf. Section 2.6). Remember that there are many flow regimes in multi-phase flow, involving two or three phases (without or with phase change), and the example shown below is for just one of those, namely two-phase dispersed flow. This regime generally consists of a topologically continuous carrier or background phase

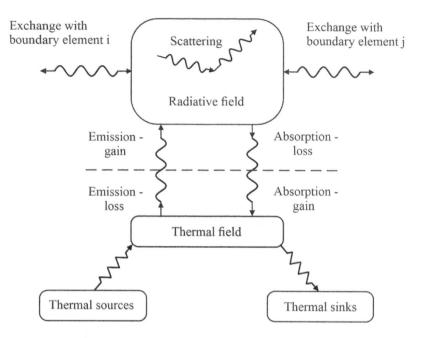

FIGURE 4.18 Schematic of the local volumetric coupling between flow (via the thermal field) and radiation, as well as boundary coupling, for a single-phase medium. A wave symbol represents radiative sources, and a resistor represents convective sources.

and a second phase consisting of discrete 'particles' which may be solid, liquid (droplets) or gas (bubbles). The term 'particle' will be used here to denote any of these three possibilities. The materials may consist of one chemical species, for example liquid water and water vapour, with or without heat transfer and phase change (evaporation or condensation), or it may involve multiple chemical species. Diabatic chemical reactions (exothermic or endothermic) may be present too. The dispersed regime can involve a gas–solid pair (solid particles in a gas), a gas–liquid pair (droplets in a gas), a liquid–solid (solid particles in a liquid), a liquid–gas pair (bubbles in liquid) and even liquid–liquid (e.g. immiscible droplets in another liquid). The present example is the gas–solid pair, with the particles being opaque to thermal radiation, and the gas is radiatively participating (see Figure 4.19). Hence, there is only one box representing radiation, but two boxes representing the thermal states of the two phases. The latter exchange heat by conduction and convection (sensible heat, and latent heat if relevant). The particles may sustain internal sources, for example, may be heated up internally by a chemical process, or may be pyrolising and generating vapour which combusts in the gas.

If there are no sources for the dispersed phase then the bottom right-hand box does not exist. That feature does appear if internal heat sources exists in the particles or if they reach boundaries, say by gravitational settling or deposition by the variety of other mechanisms which exist (such as turbulent and phoretic), since heat can then be exchanged by direct contact with the wall. That behaviour will be more

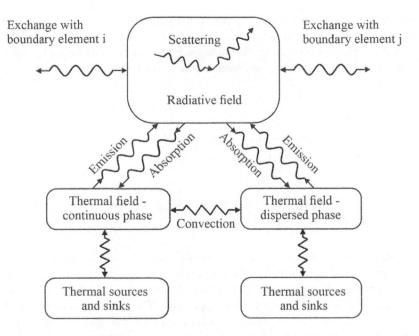

FIGURE 4.19 Schematic of the local volumetric coupling between flow (via the thermal field) and radiation, as well as boundary coupling, for a dispersed two-phase medium, with the continuous phase being radiatively participating, and the dispersed phase being opaque. A wave symbol represents radiative sources, and a resistor represents convective sources. The latter will include latent heat if phase change is occurring.

pronounced if the flow is in the dense regime (with high particle volume fraction), or if the particles' material is liquid rather than solid. However, such phenomena may also be part of a transition to another multiphase flow regime.

If the dispersed phase is not opaque, then two radiative fields will appear at the top of Figure 4.19, and they will be coupled by transmission at the interface between the phases.

The discourse here overlaps with the discussion on porous media, in Section 5.3.6.

4.4.1 COUPLING BETWEEN RADIATION AND TURBULENCE

Finally, the reader's attention is drawn to the issue of the coupling between turbulence and radiation (usually abbreviated as *TRI*).

The engineering and scientific community has of course been aware for a long time that, regarding the temperature dependence of a turbulent reacting flow, the mean reaction rate is very different to the reaction rate at the mean temperature, due to the non-linearity of the Arrhenius function, especially its exponential:

$$\text{Rate} = A \exp\left(\frac{-E_a}{RT}\right) \tag{4.63}$$

where the pre-exponential A may depend on temperature, E_a is the activation energy, and R is the universal gas constant. Similarly, the T^4 dependence of thermal radiation attracted attention, with volumetric emission being an obvious target for analysis. Decomposing temperature as usual into a mean and a fluctuation,

$$T = \bar{T} + T'; \overline{T'} = 0 \tag{4.64}$$

and initially considering just the blackbody emissive power $E_b = \sigma T^4$; expansion yields after averaging (denoted by an overbar),

$$\overline{E_b} = E_b\left(\bar{T}\right)\left(1 + 6\frac{\overline{T'^2}}{\bar{T}^2} + 4\frac{\overline{T'^3}}{\bar{T}^3} + \frac{\overline{T'^4}}{\bar{T}^4}\right) \tag{4.65}$$

As illustrated in the literature, a temperature fluctuation of ±30% will increase emission by more than 50%, although the above has ignored fluctuations in the spectroscopic properties. It appears that Spiegel and Townsend (Spiegel, 1957; Townsend, 1958) were the first to tackle this important topic. The literature indicates that it is necessary to account for this aspect if accuracy is desired, and indeed, that major errors in radiant heat fluxes are possible if this issue is ignored, varying from several tens of percent to even hundreds of percent. Early impressions were that the interactions are unimportant in non-reacting flows because fluctuations are usually much smaller than in reacting flows, and are important when reactions take place. It is possible, however, for large fluctuations to occur in non-reacting fluids at free shear layers and when those layers interact with walls. Turbulent fluctuations definitely affect radiative transfer, and moreover, radiation can affect temperature and density, and hence the flow generally. TRI is often ignored in heat transfer CFD, and one should take an holistic view of inaccuracies introduced by aspects other than radiation, but research using DNS has highlighted important effects. Modest, in his 2013 book, said 'Today the study of turbulence–radiation interactions remains an extremely active field of research'. Indeed, almost a decade later, much has been done, with DNS playing a major role in the research, and large effects of TRI being reported. It is interesting, though, that a recent paper (Guo et al., 2021) reports that one version of the WSGG model achieved relative errors for the radiation source terms of less than 10% (which is good for practical applications), for a methane jet flame, without any modelling of TRI, even if the authors say that that should be remedied. Only a little will be said below about TRI and the interested reader is referred to (Cox, 1977, 1998; Faeth, Gore, Chuech and Jeng, 1989; Kounalakis et al., 1989; Coelho, 2007, 2012; Habibi et al. 2007; Amaya et al., 2010; Poitou et al., 2012; Modest, 2021; Vicquelin et al., 2014; Silvestri et al., 2018; Rodrigues et al., 2019; Fraga et al., 2019; Armengol et al., 2020). Even in forced convection, this effect has been found to influence the classical Law of the Wall (see Figures 2.3 and 2.4) significantly.

An example of the holistic aspect is surface emissivity, which is subject to large uncertainties for some materials (e.g. rusting metals), and one needs to consider

whether the effort and cost of the more sophisticated techniques are worthwhile if
their benefits are overwhelmed by uncertainties in surface emissivity, if those uncertainties apply. DNS is not yet a practical tool for large-scale applications, but
Reynolds-averaged Navier–Stokes (RANS) and LES-based models are available in
the literature, and movement towards improved predictive fidelity will inevitably
continue, as it certainly should.

The following outline is based on (Coelho, 2007, 2012; Modest, 2021). In con-
texts other than DNS, the coupling is embodied in the time-average of the radiative
source (Equation (4.60)) in the energy equation:

$$\overline{S_R} = -\int_0^\infty \left[4\pi \overline{K_{a\lambda}I_{B\lambda}} - \int_{4\pi} \overline{K_{a\lambda}I\left(\lambda,\Omega'\right)}d\Omega' \right]d\lambda \qquad (4.66)$$

The correlation in the first term is known as the *Emission TRI*, and the one in the
second term is known as the *Absorption TRI*. The former is the one which has under-
standably received most attention; the latter is much harder to evaluate, because it
depends on fluctuations at remote locations. That was probably the main reasons for
the absorption TRI being ignored initially, but this was also rationalised using the
optically thin fluctuation approximation (OTFA), whereby

$$\overline{K_a I} \cong \overline{K_a\left(\overline{T}\right)}\,\overline{I} \qquad (4.67)$$

A Taylor expansion of $\overline{K_a I_B}$ about the mean, considering a grey medium for illustrative purposes and ignoring dependence of the absorption coefficient on species
concentrations, yields the following for the emission:

$$\pi \overline{K_a I_B} = \overline{K}\overline{T}^4 \left[1 + 6C_{TRI1}\frac{\overline{T'^2}}{\overline{T}^2} + 4C_{TRI2}\frac{\overline{T'^2}}{\overline{K_a}\overline{T}}\left(\frac{\partial \overline{K_a}}{\partial T}\right)_{\overline{T},m_i} + O\left(\frac{\overline{T'^3}}{\overline{T}^3}\right) \right] \qquad (4.68)$$

The second term in the braces is known as the *temperature self-correlation* and the
third term as the *absorption-temperature* correlation. The derivative of the absorption coefficient is at fixed composition, denoted by m_i. The constants C_{TRI1} and C_{TRI2}
were introduced by (Snegirev, 2004) for a RANS approach, to improve comparisons
with experiments, and values of 1.25 and 1.0 were recommended, respectively. The
temperature variance was computed via the standard conservation PDE for that quantity. It is reported in (Coelho, 2013) that the mean absorption coefficient is reduced
by turbulence, unlike radiation intensity, which is increased by it. Moreover, the
error in the computed emission when TRI is ignored altogether is smaller than when
only accounting for temperature fluctuations. Updated TRI modelling for RANS has
been proposed recently by (Fraga *et al.*, 2020), with the role of composition being
taken into account. Much work has also been done using LES, and (Chatterjee *et al.*,
2015) recommended $C_{TRI1} = 2.5$. Some workers have simply set $C_{TRI2} = 0$. In LES

simulations, involving filtered quantities instead of time-averaged ones, the sub-grid scale temperature variance has usually been computed as

$$\overline{T'^2} = \left(C_S\Delta\right)^2\left[\nabla T\right]^2 \tag{4.69}$$

There is also a potential for TRI to involve a feedback loop with walls; most papers have assumed a vanishing variance at the wall, but the models in (Sinai, 1986, 1987) provide a wall function for the temperature variance and predict temperature fluctuations at the wall-fluid interface of a diabatic wall.

4.5   EQUILIBRIUM OR NON-EQUILIBRIUM?

The reader will probably have noticed that the radiation literature often refers to 'equilibrium' or 'radiative equilibrium'. Several definitions of radiative equilibrium exist in the literature, and the entity discussed here is also referred to as the volumetric or pointwise radiative equilibrium. Referring to Section 4.4 (especially Figures 4.17 and 4.18), in radiative equilibrium the emission is equal to the absorption, the source term (Equation (4.60)) then vanishes, and the divergence of the radiative flux vector is zero, or equal to the internal radiative source if that exists. For pure radiation the temperature field is therefore 'floating' and determined solely by the radiative field. This is a good approximation at very high temperatures and/or when other modes of heat transfer are much weaker than radiation (see Section 4.3). As a classic grey example, the temperature T_g of a homogeneous (isothermal and constant radiative properties) plane-parallel slab bounded by diffuse black walls (at temperatures T_1 and T_2), with no additional sources, is given by

$$T_g = \left[\frac{1}{2}\left(T_1^4 + T_2^4\right)\right]^{1/4} \tag{4.70}$$

In general CFD situations the radiative state tends to be in disequilibrium, associated with the multiple coupled modes of heat transfer, which interfere with the equilibrium, and the source term is non-zero. Be that as it may, in a CFD simulation with a package possessing a coupled radiation module the software should automatically capture the equilibrium limit if that is relevant, usually when radiation overwhelms other modes of heat transfer.

Radiative equilibrium, or lack thereof, is generally a different matter to thermodynamic equilibrium, and a wide range of thermodynamic states exist in the spectrum of cases analysed by CFD. For example, the internal vibrational modes of molecules making up a gas may be in equilibrium, whilst the radiative field may be in a non-equilibrium state. See (Clarke and McChesney, 1976; Attard, 2012; De Groot and Mazur, 2013).

4.6   THE MEANING OF 'SEMI-TRANSPARENT'

A common term which is encountered in the radiation and heat transfer literature is 'semi-transparent'. Its meaning varies somewhat, and will be elaborated here.

Usually, that term is applied in the literature to solids and liquids which are not opaque over the length scales being considered. Common examples in the visible part of the spectrum are water and glass.

The semi-transparent descriptor is also used in relation to porous media, the radiative behaviour of which is discussed in Section 5.3.6. Here, the term usually refers to a porous opaque solid, the pores of which are filled with a transparent or participating fluid. The solids are sometimes themselves semi-transparent in the first meaning of the term discussed above. The morphology of the solid is general, and thus may consist of one volume which is topologically connected, or disparate objects separated from one another, perhaps in contact in places, such as fibres or packed beds.

4.7 SEMI-TRANSPARENT SLABS, WINDOWS, SOLAR RADIATION

The most familiar semi-transparent object is the common window. Windows may of course involve multiple panes of glass, most commonly in cold climes, with each often described as a 'slab' in the radiation literature. A slab involves two or more interfaces, and glass coatings (which are usually very thin).

Solar radiation is of course important in relation to the built environment, renewable energy, and the highly topical fields of meteorology and climate change. In the built environment spectral behaviour is usually categorised in terms of shortwave and longwave radiation (see Section 4.1.1). The key point is the qualitative similarity of the atmosphere and glass, with high transmissivity in the shortwave, and low transmissivity in the longwave. Relevant material may be found in heat transfer textbooks such as (Rohsenow *et al.*, 1998; Holman, 2018; Lienhard IV and Lienhard V, 2012; Modest, 2021; Incropera *et al.*, 2017; Howell *et al.*, 2021), the plentiful literature on radiation in the environment, with just a few quoted here, e.g. (Buglia, 1986; Stamnes *et al.*, 2017; Veerman *et al.*, 2021), industry guides such as (ASHRAE, 2017; CIBSE, 2021), modelling reviews such as (Sinai *et al.*, 2016), and journals such as *Energy and Buildings, Building and Environment, Energy, International Journal of Climatology, Journal of Solar Energy Engineering, Solar Energy Materials and Solar Cells*, and *Journal of Quantitative Spectroscopy and Radiative Transfer*.

The literature covers both the processes of radiation in the atmosphere as well as at and close to the Earth's surface. Here, as far as solar radiation is concerned, we will confine ourselves to two plots. Figure 4.20 shows higher-resolution reference data, which represents an average for the USA, for the 1976 U.S. Standard Atmosphere. The top curve is the extra-terrestrial irradiation, the lowest curve is essentially the direct irradiation at the Earth's surface, and the middle curve is the sum of direct, scattered and reflected irradiation at a tilted surface. The solar air mass is the ratio of the solar ray path length through the atmosphere at the given zenith angle (the angle between the sun and the local vertical), to the path length at zero zenith (i.e. sun directly above); for zeniths up to approximately $75°$ this can be approximated by the inverse of the cosine of the zenith angle. Figure 4.21 is a plot from MODIS (NASA's Moderate Resolution Imaging Spectroradiometer) which is annotated with some key bands.

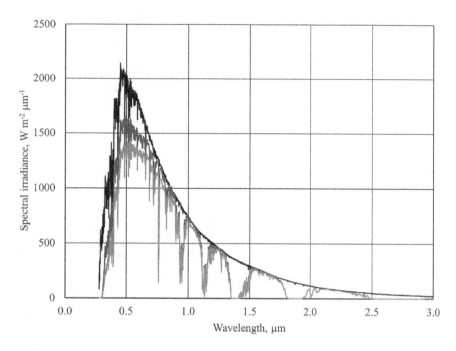

FIGURE 4.20   Example of solar irradiation. Reference Air Mass 1.5 Spectra, ASTM G-173-03, representing an annual USA average (https://www.nrel.gov/grid/solar-resource/spectra-am1.5.html). Top curve: Extra-terrestrial. Middle curve: Global tilted (37°). Bottom curve: Direct + circumsolar.

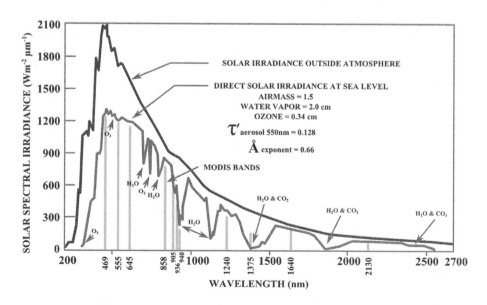

FIGURE 4.21   The MODIS measurements of solar radiation. NASA.

Glass of course plays a major role in buildings, and much work has been done on glass coatings, aiming to minimise solar heating of the building in summer, maximise solar heating in winter, and minimise radiative losses from the building in winter. 'Smart' glasses have been developed which permits control of the glass properties to suit time-varying circumstances.

Turning to simulation involving windows in the context of heating, ventilation, and air conditioning (HVAC) and energy consumption in the built environment (or in other analogous situations), the window is subject to the following radiative phenomena:

- Direct (shortwave) solar radiation.
- Scattered shortwave radiation.
- Longwave radiation arriving from the atmosphere and the surrounding ground and entities lying on it.
- Radiation arriving from the building interior.
- Absorption by the glass of some of the shortwave radiation.
- Greater absorption by the glass of the longwave radiation.
- Emission by the glass.
- Multiple reflections within panes and by adjacent glass panes in the case of multiple glazing designs.
- Refraction.

Experiments have been conducted too of course, and significant work has been done on analytical and computational models of window assemblies, mainly on a 1-D basis. See Section 4.1.9, Section 5.1 (especially Section 5.1.3), Appendix C, and Appendix K.

5 Modelling

This section deals with techniques for representing and solving for the radiation field and any other associated quantities which are required (such as radiative fluxes at boundaries or inside the computational domain). Whilst there are many elements in Chapters 2–4 which can be described as 'modelling', presently that term refers to the mathematical representation of the radiative field, which in most cases is embodied in the radiative transfer equation (RTE), and to the solution (almost always computational) of that representation.

5.1 BOUNDARY CONDITIONS

Detailed discussion of computational fluid dynamics (CFD) boundary conditions (abbreviated as BCs) is beyond the scope of this book, and plentiful information is provided by the user manuals of the CFD packages and by textbooks. However, it is essential to allocate some space here to this subject, especially when radiation is involved in the exterior region/regions, and some non-radiative aspects have to be addressed. This can begin with a list of the common categories of BCs encountered in CFD:

(a) Wall
(b) Inlet
(c) Outlet
(d) Opening/pressure boundary
(e) Symmetry plane
(f) Porous/permeable wall

The issue of the spatial extent of domain is important of course, and was only mentioned cursorily in Chapter 3. A simple illustrative example from the built environment will be outlined here in order both to elaborate a little on the CFD aspect, and to support the discussion on external radiative boundary conditions. The situation reflects many scenarios facing CFD practitioners when deciding the extent of the CFD domain. The example involves a building with open doors, windows or vents (Figure 5.1). In fact, the majority of buildings do leak even when all such features are closed, which affects ventilation, indoor air quality and energy efficiency, e.g. (CIBSE, 2000, 2021; ASHRAE, 2017). Leakages can also have a large impact on internal fire dynamics (Sinai, 1999), wherein the pressure inside the building during the event can at times be larger or smaller than that in the external ambient. In the built environment sector leakages are also known as *infiltration* and *exfiltration*.

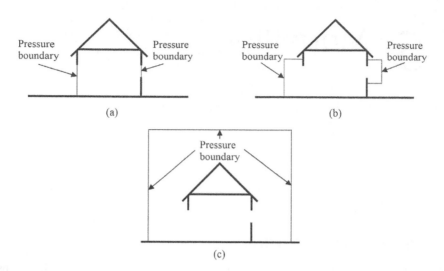

FIGURE 5.1 Example of pressure boundary positioning, at a building with open doors, windows, or vents. Slice through the idealised building, door on the left, window on the right.

The example features 'openings' or 'pressure boundaries'. That is a useful utility, which determines locally what the 'appetite' of the local flow is; if the pressure just inside the boundary is greater than the pressure just outside it, then fluid will leave through the boundary, and the converse applies, with ambient fluid (possessing user-specified properties) entering the domain if the internal pressure is lower than the ambient. Thus, fluid may leave the domain over some parts of the boundary, and ambient fluid may enter the domain over other parts of the same boundary, all automatically determined by the code.

In option (a) in Figure 5.1, the outer computational boundary is flush with the door and window. This is reasonable if the door and window are closed. However, if they are open, placing the pressure boundaries at apertures is poor practice, because pressure boundaries usually attempt to impose a small or vanishing normal pressure gradient, yet pressure gradients are large at the aperture edges, and pressure loss for the flow passing through the aperture is poorly predicted. Conversely, the flow rate through the aperture is poorly predicted for the given driving head, whether driven by wind or gravity forces. Option (b) will do much better, and is the least which should be done. The best is option (c).

The sketch shown here applies to a nominally stagnant ambient/atmosphere, and the building is subject to *natural ventilation*. If wind exists, the pressure boundary/boundaries need to be much further from the building/buildings. Indeed, it is worthwhile to mention the terms used for the very different spatial extents of CFD models in the built environment and meteorology, and thermal radiation aspects therein. Indoor models are typically tens of metres in scale, and radiation modelling is not uncommon, although non-participating approaches tend to be used. Building scales are of the order of a few hundred metres, and take one to the realm of computational wind engineering (CWE). Microscale models are up to about 2 km in size, typically involve several buildings within a town or city (or a number of wind turbines in wind farm), and usually do not involve volumetric radiation. Atmospheric stability is considered here as well as at larger scales, with CFD models applied to vertically stable, neutral and unstable atmospheres, e.g. (Jacobson, 2005; Turner, 2012). These models can be used to represent urban microclimates, as can the next scale. Mesoscale entails a whole city and some of its surroundings, of the order of a few hundred kilometres, and takes one towards the realms of atmospheric and geophysical fluid dynamics (including heat transfer and dispersion), and meteorology. Volumetric radiation is included sometimes. Meteorological scales involve countries, continents, or the whole globe, and volumetric radiation is an important aspect thereof. A range of modelling approaches is exemplified by (Haupt et al., 2019 and Walters et al., 2019).

Best-practice guidelines are available. For example, if wind is approaching from the left, then the left vertical boundary in Figure 5.1c will usually be an inlet, with an imposed power-law velocity profile, or preferably in the author's view, a logarithmic rough-wall profile with consistent profiles of turbulence quantities, e.g. (Franke *et al.*, 2011). The specified profile is usually compatible with an upstream turbulence field which is in equilibrium, the ground aerodynamic roughness upstream of the inlet, and a prescribed wind speed and direction at a height of 10 metres. Typically, radiation effects on the microscale profile are ignored. All three outer boundaries (and lateral boundaries in 3-D simulations) would then need to be much further than the building. At least some of the adjacent buildings need to be modelled, in line with the microscale approach, and powerful computers have been used to model urban flow over entire towns or cities, at varying levels of detail.

At openings it is essential to specify the boundary pressure accurately. A pitfall which can be encountered occurs when inadequate account is taken of hydrostatic effects in buoyant flow. Thus, even if an error of only a few Pascals is imposed it can, for example, lead to very strong and false currents at the two top corners in Figure 5.1c, which can affect the whole domain.

This brief discussion illustrates some of the issues which need to be taken into account when designing the domain, incorporating additional factors besides radiation. Attention will now turn to the radiation aspects.

There are two aspects to radiative BCs: A boundary condition imposed on the intensity if the radiation algorithm is solving for that entity inside the computational domain, and BCs representing phenomena beyond the external boundary of the computational domain. The former is not one which the practitioner would normally alter directly, other than by choosing modelling and properties options offered by the software, and may be found in CFD user manuals and the textbooks already cited. The latter is a different matter and requires the practitioner's attention.

5.1.1 Opaque Walls

An illustrative sketch of the situation at an opaque wall boundary is shown in Figure 5.2. The computational domain is labelled 'Interior', and is the region to the right of the CFD boundary 'B'. In this particular example, immediately to the left of B is a masonry wall (between 'B' and 'C'), and to its left is the atmosphere.

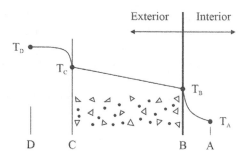

FIGURE 5.2 Schematic of the temperature profile at an opaque CFD boundary in a steady situation. In this example, B is the outer boundary of the CFD domain, which is on the right. The exterior lies to the left of B, and consists of a masonry wall and some of the atmosphere beyond.

Everything beyond B, in this case the wall and the atmosphere, is labelled 'Exterior'. Analogous situations exist in many other heat transfer applications.

It should be pointed out here that there may be reasons for extending the computational domain to the left of B. The case for doing so is strong when a simple steady-state 1-D model of the boundary conditions, as discussed in this section, is known to be inaccurate and to have a significant and harmful impact on the solution reliability. For example, the 1-D approach is poor if significant temperature gradients exist in the plane of BC. Some codes permit a halfway house in which BC is modelled as a shell, allowing such transverse gradients to be captured. If necessary, one can go further and resolve the temperature in all directions with BC being defined as a CHT region, and ending the domain at C or beyond; this is discussed below.

A key factor before that step is taken is whether the problem is steady or transient. If it is transient, then a steady-state model of B to D could be very inaccurate. The transient behaviour of the wall, and thermal inertia, may need to be accounted for, say by extending the model to C, or constructing a user routine, or coupling to another software package which analyses the walls only. In terms of the electrical analogy, a simple approach is to represent the thermal inertia as a capacitor, as is done in zone models. Having said that, for illustrative purposes, the description now continues by assuming steady conditions.

A temperature profile is shown too in the figure, in the case where heat is flowing from the exterior to the interior. Fluid boundary layers exist immediately to the right of B and immediately to the left of C. Point 'A' represent the first node next to the wall, and its role will depend on whether the practitioner has instructed the software to resolve the boundary layer or to use wall functions (which are discussed in the software manuals and fluid dynamics textbooks). Whatever the practitioner has decided in that regard, that aspect is generally handled internally by the software, and it is everything to the left of B which needs to be specified by the practitioner one way or another as a boundary condition. This is a key difference between CFD and zone models. The latter do not resolve any of the boundary layers, and do impose the

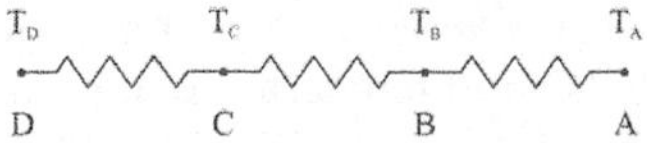

FIGURE 5.3 Electrical analogy of the wall boundary example. T is temperature. Resistance is the inverse of the heat transfer coefficient. In zone models three resistances connect the interior and exterior. In the CFD analysis, two resistances are usually employed for the boundary condition, between B and D.

boundary condition all the way from A to D; in the built environment that is embodied in the so-called 'U-value', which is essentially the effective heat transfer coefficient between A and D. In contrast, the CFD code resolves the boundary layer at B (and at C if the model is extended). Another good example of the extended approach is a heat exchanger, with the CFD code able to analyse flow of one fluid tube-side, another fluid shell-side, and conduction in the tube walls.

A steady-state representation of the current example is shown in Figure 5.3, in terms of its electrical analogy.

By default, most CFD packages expect a linear thermal boundary condition to be imposed at B, i.e. a linear relationship between the heat flux q_B at B and the temperature T_B at the same location. Its generic form can be written as

$$\hat{A}q_B + \hat{B}T_B + \hat{C} = 0 \tag{5.1}$$

This covers all of the following cases:

i.   Specified temperature T_w ($\hat{A} = 0$, $\hat{C}/\hat{B} = -T_w$). Mathematically, this is known as a Dirichlet condition.
ii. Adiabatic wall ($\hat{A} = 1$, $\hat{B} = 0 = \hat{C}$). This is known as a Neumann condition.
iii. Specified non-zero heat flux q_w ($\hat{B} = 0$, $\hat{C}/\hat{A} = -q_w$). This is also known as a Neumann condition.
iv. Specified heat transfer coefficient h to a source/sink temperature T_D ($\hat{A} = 1$, $\hat{B} = h$, $\hat{C} = -hT_D$). This is known as a mixed, Cauchy or Robin condition.

The adiabatic case merits special mention. If a boundary is well insulated say, then it is the total heat flux which is zero, and not its individual components. Thus, for such surfaces the net radiative flux equals the convective flux in magnitude, but is opposite in sign, and the surface temperature is a floating quantity, which depends on the balance between convection and radiation. The convention in this book is a positive heat flux when heat is being transferred into the computational domain. Referring to Figure (4.8) and Equation (4.16), and working in terms of total quantities, the temperature of an adiabatic grey surface may be estimated by setting the abovementioned sum to be zero:

$$\varepsilon\left(G_w - E_w\right) - h_w\left(T_w - T_g\right) = 0 \tag{5.2}$$

where h_w is the fluid-side convective heat transfer coefficient (i.e. between A and B in Figure 5.2), T_g is a true bulk fluid temperature adjacent to the wall, and E_w is the wall blackbody emissive power. In terms of the wall surface temperature this is a quartic of the form

$$T_w^4 + aT_w + b = 0 \qquad (5.3)$$

where a and b are positive and negative, respectively. Analytical solutions are possible but the relevant root can easily be found numerically. The CFD software computes this temperature as part of the solution, but these equations are described here in order to highlight the behaviour at insulated boundaries. If the magnitude of the irradiation is smaller than that of the wall blackbody emission then the wall temperature will be lower than the bulk fluid temperature, and the converse applies if the irradiation is larger than the blackbody emission. If convection is negligible, then the net radiation is zero, and the surface is described in the literature as 'reradiating'. A reversed process is used when employing black globe thermometers to measure the Mean Radiant Temperature (see Section 4.1.12). The irradiation arriving at the globe is expressed as σT_R^4, and the 'wall' (i.e. globe) temperature is measured, whence the MRT can be deduced directly, provided the true fluid temperature near the device is measured too:

$$T_R = \left[T_w^4 + \frac{h_w}{\sigma\varepsilon}\left(T_w - T_g\right) \right]^{\frac{1}{4}} \qquad (5.4)$$

Sinai et al. (2016) describes how combination rules can be used to estimate the heat transfer coefficients to allow for transitions at the globe between laminar and turbulent flows as well as between natural and forced convection.This discussion is related to the important matter of 'contamination' of temperature measurements in a rig or in the field when thermal radiation is present, e,g, Holman (2018). In such situations, embodied by the first of the two terms in Equation (5.2), it is advisable to employ shielded thermocouples, even if they are more expensive than the unshielded variety. The error can be relatively large even at ambient temperatures, and can be hundreds of degrees Celsius in fire scenarios, mainly in cold regions exposed to a nearby flame or warm smoke products. If the thermocouples are unshielded, then it is essential to estimate the error, by measuring or calculating the incident radiation. Thus, if a CFD analysis is used to validate software modelling, then one of the following is advised, applied to Equation (5.2) together with empirical correlations for the convective HTC:

a. Experimental approach: Use the incident radiation, measured or derived from the CFD analysis (or some other theoretical model), to correct the measured temperature. This deduces the true fluid temperature.
b. Theoretical approach, to compute the 'contaminated' temperature: Model the thermocouple either in detail or using empirical correlations to estimate

the error, using the computed incident radiation at the thermocouple, and add that to the computed fluid temperature in order to simulate the thermocouple's readings.

Now consider (iv). As discussed, the heat transfer coefficient h (abbreviated as HTC) covers phenomena between B and D. Assuming initially that in any of the regions in the exterior only a single mode of heat transfer exists (say convection), then h (or h_{BD}) is given by the classical results for resistances in series:

$$\frac{1}{h} = \frac{1}{h_{BC}} + \frac{1}{h_{CD}} \qquad (5.5)$$

The subscripts 'BC' and 'CD' have a clear meaning. In this example, simple conduction would be used for BC, and a correlation for CD. That correlation can be used to estimate a prescribed and constant value of the HTC, or in a more accurate approach, to account for the temperature difference between C and D, which would usually require an iterative calculation of h and the temperature at the interface C in this example (as a function of T_B). Here are two simple cases for the present example:

If a simple approximation of h_{CD} is a constant, then, for constant wall conductivity,

$$h = \left(\frac{L_{BC}}{k_{BC}} + \frac{1}{h_{CD}}\right)^{-1} \qquad (5.6)$$

where L_{BC} and k_{BC} are the wall thickness and conductivity respectively. If the external fluid is nominally stagnant, and heat transfer outside C occurs only by natural convection in the turbulent regime, then the correlation $Nu = CRa^{1/3}$, e.g. (Holman, 2018), is suitable, where Nu is the Nusselt number, Ra is the Rayleigh number (cf. Appendix A), and $C \cong 0.1$. The heat transfer coefficient can then be estimated by

$$h = \left\{\frac{L_{BC}}{k_{BC}} + \frac{1}{k_D}\left[\frac{0.5\nu\lambda\left(T_C + T_D\right)}{gC|T_D - T_C|}\right]^{1/3}\right\}^{-1} \qquad (5.7)$$

where k_D is the conductivity of the external fluid, ν is that fluid's kinematic viscosity, and λ is the molecular thermal diffusivity. In some cases, account needs to be taken of variations of k_D with temperature. The appearance of T_C requires iteration, which is possible in some codes as an inbuilt facility, with a full non-linear capability. If that is not available, then a first approximation simply replaces the arithmetic mean in the numerator of the second term by T_D, and in any case that is often used as the reference temperature to define the Rayleigh or Grashof number. The next idealisation concerns the denominator in that term, and case-specific engineering judgement is required for that. In some cases, $|T_D - T_C|$ is well approximated by $|T_D - T_B|$.

All this is of course readily generalised for an arbitrary number of regions, larger than the two considered in the present example. A number of numerical approaches are available in the literature for computing the heat flux across multiple regions with non-linear heat-flux laws in each. The common approach, using the HTC as in Equation (5.5), leads to an iterative process, essentially representing $N - 1$ coupled transcendental equations for the node/interface temperatures, where N is the number of resistances. This is normally straightforward but can be challenging for highly non-linear phenomena such as boiling, and a simpler approach formulates the problem in terms of the heat flux itself instead of individual temperature drops (Adiutori, 1965). Thus, if the flux law for N resistances, expressed in terms of the HTC (simply by dividing the heat flux by the temperature difference) is of the form

$$h_i = C_i \left(\Delta T_i \right)^{n_i} \tag{5.8}$$

then summing the temperature drop across all the elements gives

$$\sum_{i=1}^{N} \left(\frac{q}{C_i} \right)^{\frac{1}{n_i+1}} - \Delta T_{\text{overall}} = 0 \tag{5.9}$$

where $\Delta T_{\text{overall}}$ is the total temperature drop across the assembly. Usually, the index n_i (not to be confused with the refractive index) is 0 for forced convection, in the range 1/4 to 1/3 for natural convection, and zero for simple conduction. Equation (5.9) is a single transcendental equation, and q can be determined numerically as its root. For combinations of series and parallel elements, the contribution of any of the parallel components making up a parallel element will do, since once the total flux is determined then the temperature drop across the parallel element is known and fluxes through the remaining parallel components in the given element follow.

Given the relatively high frequency of occurrences of CFD situations in which empirical correlations are involved in boundary conditions (and other parts of the domain), a little more will be said here about a few common correlations. Advances in computing power means that CFD practitioners can afford to be increasingly ambitious, and extend the domain further to avoid the correlations, but there is usually a balance to be struck.

Two transitions in the Nusselt number will be addressed here (Appendix A). First, in order to cover the transition between laminar and turbulent flows, a simple approach adopts the larger of the laminar and turbulent values, i.e.

$$Nu_{\ell t} = \max \left(Nu_\ell, Nu_t \right) \tag{5.10}$$

The second transition is between forced and natural convection, and is known as *mixed convection*. A combination rule, commonly named after S.W. Churchill, is

$$Nu = \left(Nu_{\ell tF}^n + Nu_{\ell tC}^n \right)^{1/n} \tag{5.11}$$

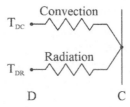

FIGURE 5.4 Electrical analogy of multimode heat transfer (convection and radiation) between C and D, to different source temperatures T_{DC} and T_{DR}.

where typically $n = 3$, and the subscripts F and C denote forced and natural convection. Thus, four correlations are synthesised into one continuous correlation which is convenient for computations. One familiar convective correlation is the one which has already been cited above, for turbulent natural convection. Another which will be cited here is the so-called Dittus–Boelter correlation for turbulent forced convection:

$$Nu_{tF} = 0.023 \, Re^{0.8} \, Pr^{0.3} \tag{5.12}$$

This was originally developed for internal flow in circular tubes (with small to moderate temperature difference between the tube surface and the centreline fluid), but has sometimes been found to be realistic for other configurations too.

Returning to the current example (Figure 5.2), we allow for both convection and radiation to exist between C and D. The electrical analogue now consists of two resistances in parallel (Figure 5.4). If only one temperature exists ion D, the HTC for region CD is simply given by

$$h_{CD} = h_{CDC} + h_{CDR} \tag{5.13}$$

where the subscripts 'C' and 'R' refer to the convective and radiative components respectively. This is no different to simply adding the two fluxes. However, in some cases one may wish to impose different source temperatures for each heat transfer mode, denoted here by T_{DC} and T_{DR}, respectively.

The fluxes can simply be added. If the software insists on a Cauchy condition with a single HTC to a single temperature, then using Equation (5.13), the linear heat transfer law for CD can be written as a single HTC to a fictitious, weighted source temperature T_D' as follows:

$$q_{CD} = h_{CD} \left(T_D' - T_C \right) \tag{5.14}$$

where

$$T_D' = \frac{h_{CDC} T_{DC} + h_{CDR} T_{DR}}{h_{CDC} + h_{CDR}} \tag{5.15}$$

As an illustration, assume that the radiative component is well described by the following common form:

$$q_{CDR} = \varepsilon_C \sigma \left(T_{DR}^4 - T_C^4 \right) \tag{5.16}$$

where ε_C is the wall total emissivity at C. As already mentioned, some codes permit non-linear heat laws, sometimes via user routines for those purposes. This is particularly relevant to radiation because of the highly non-linear T^4 trends. If the code does not offer this option, then this is typically linearised as

$$h_{CDr} = 4\varepsilon_C \sigma T_m^3 \tag{5.17}$$

where T_m is a mean temperature, which can be approximated by T_{DR}, T_C, or the arithmetic mean of those two.

It is emphasised again that this discussion relates to the external boundary condition, rather than the internal condition at a boundary; the latter in its entirety is normally handled internally by the software.

5.1.2 COLLIMATED RADIATION

Some CFD packages allow the user to specify a collimated source at a boundary, although that facility is usually restricted to a subset of the radiation solvers offered by the code. Be that as it may, that is how collimated radiation can be imposed at a wall with all the associated phenomena at a wall such was friction and convective heat transfer, and this can mimic a window.

5.1.3 SEMI-TRANSPARENT BOUNDARIES (WINDOWS)

The subject of windows (also known as fenestrations) was introduced in Section 4.7, and is elaborated on here. The CFD practitioner has several options for modelling the windows:

i. This option is illustrated in Figure 5.5. When meshing the building interior (A in the figure), end the domain at the inner surface of the window assembly, and model the window and external environment (B in the figure) in terms of a boundary condition, both for convective and radiative heat transfer, employing an analytical or numerical model of the fenestration. This would normally account for the (shortwave) direct and diffuse solar radiation, and the general longwave radiation. Two radiative sources are shown arriving at the interface from the exterior (environment). The subscript R denotes radiative sources, E denotes exterior, C denotes collimated direct solar radiation, and D denotes diffuse radiation, made up of both scattered solar (shortwave) radiation and diffuse longwave radiation. It is possible to run this model on a grey basis. In a spectral CFD model the diffuse sources

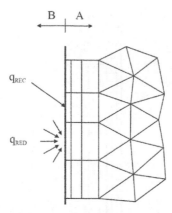

FIGURE 5.5 Coarse-meshed sketch for Model (i) of radiative boundary conditions at a window. The building interior is in region A, and the thick vertical line is the interface between the interior and the interior face of the first pane of glass. B is the region beyond that interface, including fenestration, and is not modelled explicitly in the CFD analysis.

would be separated. The values of the imposed radiative fluxes should account for attenuation and transmission of the ambient fields by the fenestration (windows), discussed below. Thermal inertia of the fenestration would be a challenge in transient cases.

This model is equivalent to option (a) in the discussion of Figure 5.1.

ii. Extend the domain to include the glass pane/panes (and any air gaps which exist), with the panes meshed up and modelled as CHT objects (Figure 5.6). This avoids the need for a sub-grid model of the fenestration, enables fenestration design with CFD, and improves the modelling for both steady and transient problems. The boundary conditions reflect the actual ambient state.

In Figure 5.6 the mesh within the fenestration does not match the mesh in the interior fluid (A); some packages permit that, but if the package does not, then cell faces in A need to match cell faces in B (at the A–B interface).

iii. Extend the domain further to include some of the external atmosphere. However, exclude the window assembly, and simply model the assembly as an interface between the interior and exterior environment, with a crude representation of the window behaviour. As already illustrated in the discussion about Figure 5.1, the positioning of boundaries is influenced by a number of factors, not only involving radiation. From the radiative point of view, this setup is similar to Model (i), in that the user is implementing some sort of sub-model of the fenestration in both cases, but it does, of course, differ from the flow viewpoint. Meshing both a building interior and exterior is wise for situations such as natural ventilation, and improves calculation of conditions at the building exterior (and hence interior) even if the building envelope is impermeable (because of heat transfer) (Figures 5.7 and 5.8).

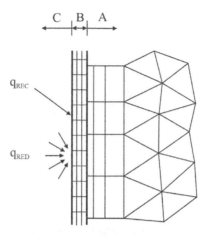

FIGURE 5.6 Coarse-meshed sketch for Model (ii) of radiative boundary conditions at a window. The building interior is in region A, and region B is the fenestration, which may be one pane of glass, or several panes and air gaps, and a thermally conducting frame. C is the external region beyond the fenestration.

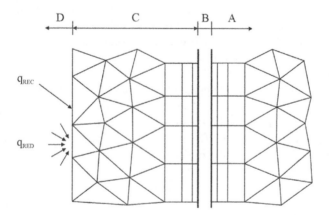

FIGURE 5.7 Coarse-meshed sketch for Model (iii) of radiative boundary conditions at a window. The building interior is in region A. Region B, which is excluded from the computational domain, is the fenestration, which may be one pane of glass, or several panes and air gaps. C is the external region beyond the fenestration, and D is the ambient outside the computational domain.

iv. Extend the domain further than (ii) to include some of the external atmosphere as in (iii). But in this case, as in (ii), include the pane CHT items, and spectral radiation. This is the most comprehensive option.

It should be remembered that when an external radiative source is to be imposed, as happens in the built environment, then moving the boundary away from the window, as in (iii) and (iv), requires a spectral model to ensure that attenuation between

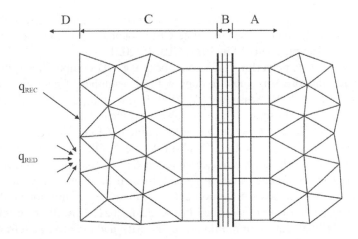

FIGURE 5.8 Coarse-meshed sketch for Model (iv).

the boundary and the window is not over-predicted, as would generally happen with a grey model. Attenuation of radiation emitted by fire, e.g. (Fuss and Hamins, 2002), involves larger wavelengths than solar radiation. However, the external sources could be imposed at the outer boundary of the fenestration (the B-C interface) for models (iii) and (iv).

It should also be noted that if the glazing is included explicitly, it will consume a large number of computational cells if the building is not small, because of the large number of windows, the small thickness of the panes compared with their other two dimensions, and the additional resource required for double and triple glazing. Moreover, coatings are not uncommon, and these are very thin, whereupon resolving the coatings on every pane in a large building can render the computational effort prohibitive or at least questionable. One way in which this can be tackled is a detailed analysis of a single window assembly, with coatings if necessary, in order to derive an idealised 1-D model, or a lookup table, which can be deployed at any location.

Evidently, much depends on the data available to the practitioner about the properties of the window assembly and its constituent parts. Zone models of heating, ventilation, and air conditioning (HVAC) typically use the so-called U-Values (or U-Factors) for the windows (and other parts of the building envelope), and the CFD modeller could take the same approach, although CFD is capable of analysing windows more accurately. The U-Value approach does not resolve the convective and radiative components accurately in the general sense, for example when the 1-D assumption is inappropriate.

It should also be noted again that the U-Value includes the resistance of the boundary layer at the inner surface of the window (or general building) envelope. CFD software, on the other hand, usually calculates that resistance, and expects to be provided with the resistance of everything beyond that surface, i.e. the window-pane or panes, gaps, and the external boundary layer. This is an important matter not to be ignored.

Much information on window modelling, not necessarily in the direct context of CFD but relevant to models used to define CFD boundary conditions can be found elsewhere, for example (ASHRAE, 2017; CIBSE, 2021).

Be that as it may, powerful, standalone, bespoke 1-D computational models of windows are available, some able to cope with any number of panes and coatings (e.g. Crawley *et al.*, 2004; Arasteh *et al.*, 2009; Lyons *et al.*, 2010), and these could be coupled to the CFD setup. There are also analytical 1-D models available for single and double glazing; these are easier to couple to CFD setups.

The author's opinion is that if a 1-D sub-grid model of the window would suffice, then the ideal approach would use experimental data if coatings are involved, and would preferably be detailed enough to break down the convective and radiative characteristics. If such data are not available or cannot be commissioned, then analytical models such as those described in (Modest, 2021; Howell *et al.*, 2021) for radiation, and (Sinai, 2003) for multimode convective and radiative heat transfer are appropriate. Alternatively, and probably more accurately, one of the bespoke computational packages already mentioned can be coupled to the CFD.

If the window components are to be represented explicitly, then the user will need to specify the refractive indices of the materials, and their absorption coefficients (which will usually be spectral).

Having emphasised the spectral behaviour of glass, and the likely requirement to set up a spectral CFD model, it is incumbent to point out that some simulations involving solar heating can be accomplished with a grey model. For scenario (i) above, the direct shortwave radiation passing through a window at the domain boundary can simply be represented as a collimated source (if the software permits that). Additional sources such as the diffuse shortwave external field, and the diffuse longwave external field, can all be accommodated in whatever model has been set up at the boundary, all within a grey framework, after calculating the transmission through the fenestration.

Sometimes the behaviour of windows is also included in the CFD simulations of fires. The collapse of windows during a fire, besides being a hazard as of itself, is of course likely to have a major impact on the fire dynamics and smoke movement. Such an event can be caused by two mechanisms, both of which may be directly or indirectly related to the thermal radiation:

a) Pressure loading across the windowpanes (which can be remote from the fire). Pressure differences between the interior and exterior of a building on fire can be hundreds or thousands of Pa. The pressure difference can be positive or negative during the fire evolution (Sinai, 1999).

b) Cracking due to thermal stresses and compression in the plane of the sheet by the window frame as the glass is heated by the fire. That heating may be by radiation only, or radiation and convection, depending on the shape and location of the flame and hot products.

Such simulations may involve one-way or two-way coupling (thermal but can be mechanical too) between the CFD software and structural analysis software.

5.1.4 INLETS, OUTLETS AND OPENINGS

The names of 'inlets' and 'outlets' are self-explanatory. They usually involved specification of velocity or mass flow rate, and at inlets also entities like temperature turbulence, and concentration.

The term 'opening' is sometimes interchanged with 'pressure boundary'.

At inlets, outlets and openings the radiation incident on the boundary from the interior is removed from the field without being reflected. Typically, the user is given two options for radiation delivered to the boundary from the exterior:

- Diffuse blackbody radiation at the boundary temperature. The latter may be part of the solution rather than being prescribed a priori.
- Diffuse blackbody radiation at a specified external temperature.

Some codes permit further radiative sources at these boundaries, including collimated radiation.

5.1.5 SYMMETRY PLANES

Symmetry planes are treated by the radiation solvers as specular surfaces.

5.2 INITIAL CONDITIONS

For transient cases, CFD packages typically allow the user to specify a homogeneous isotropic intensity field, or one based on blackbody radiation at a given temperature. Alternatively, the initial conditions may be based on a steady state which has already been simulated.

5.3 SPECTROSCOPIC PROPERTIES

The term 'spectroscopic' refers to the material's properties which affect the thermal radiation field. In other words, the absorption and scattering coefficients and the refractive index.

The issue of spectroscopic properties in various media is a wide-ranging and important subject. Since this book aims at thermal radiation in the context of CFD, the focus in this chapter will be on infrared radiative properties of gases, with a nod to liquids, soot, and glass (because the last three are sometimes involved in CFD simulations). This chapter also targets phenomena at moderate temperatures up to those typical of hydrocarbon combustion and common aspects of engineering, and ignores ionisation regimes. Representative literature may be found in (Tien, 1968; Edwards, 1976; Sparrow and Cess, 1978; Ozisik, 1985; Brewster, 1992; Goody and Yung, 1995; Lallemant *et al.*, 1996; Rohsenow *et al.*, 1998; Viskanta, 2008; Demarco-Bull, 2012; Wendisch and Yang, 2012; Modest, 2021; Howell *et al.*, 2021), and papers in publications such as the Journal of Quantitative Spectroscopy and Radiative Transfer. The material here draws on the literature and especially (Modest, 2021; Howell *et al.*, 2021), but detailed discussion is beyond the scope of this book.

The three modes of radiative exchange, already introduced above, are absorption, emission, and scattering. For most engineering application, gases only affect absorption and emission, and scattering is only significant in the presence of particulates such as dust, soot, and water droplets. Information on molecular scattering may be found in the literature on molecular physics. Particulates generally absorb and emit too, as happens in a pulverised coal-fired boiler. At the other extreme, a sodium aerosol cloud which can appear in the cover gas of a liquid metal (e.g. sodium) breeder nuclear reactor behaves almost as a pure scatterer. For information on particulate scattering the reader is referred to Section 4.1.10 and to (van de Hulst, 1981; DiMarzio, 2011) for example, where among other issues the classical Lorenz–Mie–Debye theory is detailed.

Consider gases first. The translational kinetic energy of atoms and molecules is manifested in the medium's temperature, and gas thermal radiation is associated with transitions of internal energy modes of vibration and rotation, and electronic states. Non-polar gases, i.e. those not exhibiting an electric multipole due to separation of electric charge, such as O_2 and N_2, do not emit thermal radiation and are essentially transparent to incident thermal radiation. However, polar molecules such as CO_2, H_2O, and hydrocarbon gases absorb and emit over a wide temperature range. CO_2 and H_2O are of course key greenhouse substances.

When an EM (electromagnetic) wave or photon interacts with a gas molecule, scattering is negligible as regards heat transfer, due to the smallness of the molecule (Mie theory indicates that scattering is proportional to the fourth power of the particle size), and the process is governed by absorption and emission, corresponding to an increase and decrease respectively in the molecule's energy. Three types of the molecule's energy transition exist: (a) "Bound–Bound", in which both states of the molecule are undissociated. (b) "Bound–Free" or "Free–Bound", involving transition from a bound state to a dissociated state or vice versa. (c) "Free–Free", in which both states are dissociated.

Bound–Free and Free–Free transitions pertain to high temperatures, when dissociation and ionisation are significant, and their presence is manifested at short wavelengths, from ultraviolet (UV) to visible. At the more moderate temperatures typical of most engineering applications, dissociation and ionisation are usually negligible, and Bound–Bound transitions are dominant. The literature does however suggest that dissociation can play a non-negligible role in the behaviour of turbojet combustion chambers, which as stated in the Introduction, operate at about 2000°C. At typical combustion temperatures the emission peaks in the near-infrared (0.7 to 2.5 microns) and lower end of the mid-infrared (2.5 to 10 microns). At temperatures of the order of 2000°C or higher, (b) and (c) above add a continuous element to the radiation, and noticeable effects on heat transfer and flow, although (Goody and Yung, 1995) suggests that this can occur at temperatures as low as 1000 K.

Due to the quantisation of energy states (such as electron rotation about a nucleus, linear vibration of atoms within a molecule, or rotation of atoms within a molecule), the photons involved in Bound–Bound transitions of the molecules with which they interact must possess discrete wavelengths or frequencies in order to be absorbed or emitted, leading to an infinite set of discrete spectral lines for absorption or emission (the so-called eigenfrequencies and associated eigenfunctions). Furthermore, due to various mechanism, these spectral lines are broadened slightly, and instead of being monochromatic are slightly polychromatic; much material on this and the related issues exists in the cited literature.

Generally, spectral lines between UV and near-infrared (10^{-2} to approximately 2 microns) correspond to a change in an electron orbit, those between 2 and 20 microns correspond to molecular vibration disturbed by rotational vibration, and those above 20 microns (the far infrared) correspond to molecular rotation. The vibration–rotation behaviour leads to a large number of closely spaced lines which can overlap due to line broadening. Water vapour and carbon dioxide usually dominate gaseous radiation in typical combustion applications (although fuels and associated species may be important too at times), with the water vapour primary infrared vibration–rotation bands at approximately 1.4, 1.9, 2.7, 6.3 and the rotation band above 20 microns, and with CO_2 vibration–rotation bands at 2.0, 2.7, 4.3, 9.4, 10.4 and 15 microns. For greater detail than the literature already cited the reader is referred to the sources on quantum mechanics and spectroscopy, e.g. (Penner, 1959; Hollas, 2004; Steinfeld, 2012; Griffiths and Schroeter, 2018). Interestingly, uncertainty about aspects of water vapour was identified relatively recently, with implications for global climate predictions and remote sensing, witness the developments regarding the so-called water vapour continuum involving absorption windows between the visible and microwave, within the rotational and vibrational–rotational bands (Shine *et al.*, 2012; Radel *et al.*, 2015; Shine *et al.*, 2016).

The radiative properties of gases vary strongly across the frequency spectrum, and generally the grey assumption is not a good one for gases. However, the grey approach should not automatically be dismissed and can, on occasion, produce results of reasonable quality.

The contribution of soot is very different in character, with a much smoother dependence on wavelength than that of gases. The behaviour depends on the ratio of the particle size to the electromagnetic wavelength, and whilst soot results in absorption, emission, and scattering, the latter varies like the inverse of wavelength when the above ratio is small and can be ignored for very small particles. However, soot particles can grow substantially in size due to agglomeration. Further details may be found in Section 5.3.7.

Some bands typical of combustion situations are given in Table 5.1.

Attention will now turn to the modelling of radiative properties.

5.3.1 Overview

Many approaches and approximations to radiative properties exist. The focus here will be on gases as well as gas–particle mixtures.

An important observation is made here that some property models do not deliver the local absorption coefficient, which is the parameter normally required by the radiation solver in CFD software, but rather the gas emissivity. The gas / fluid emissivity is typically defined in terms of partial pressures of chemical species, the mixture temperature, and a 'mean beam length' (denoted here by L_{MB} or by the acronym MBL). L_{MB} has been evaluated and documented for a number of generic shapes filled with a homogeneous absorbing-emitting medium. The challenge facing the CFD

TABLE 5.1

Typical Molecular Bands in Combustion (Grosshandler, 1993; Howell et al., 2021)

Species	Band (µm)
CO_2	2.0
CO_2	2.7
CO_2	4.3
CO_2	9.4
CO_2	10.4
CO_2	15.0
H_2O	1.38
H_2O	1.88
H_2O	2.7
H_2O	6.3
H_2O	20–∞
CO	2.35
CO	4.67
CH_4	1.71
CH_4	2.37
CH_4	3.31
CH_4	7.66
Soot	0.4–2000

practitioner, when using charts or formulae for the medium's emissivity (or transmissivity τ) in a grey simulation, is to translate this information to an absorption coefficient. One simple option is to assume a value for L_{MB} and deduce the absorption coefficient K_a from the Beer law:

$$\varepsilon_g = 1 - \exp\left(-K_a L_{MB}\right) \qquad (5.18)$$

or

$$K_a = -\frac{1}{L_{MB}} \ln\left(1 - \varepsilon_g\right) = -\frac{1}{L_{MB}} \ln\left(\tau_g\right) \qquad (5.19)$$

The weakness of this approach is obvious, unless the overall domain shape is a simple one and conditions therein are homogeneous, in which case the mean beam length is likely to be well defined. Some CFD packages allow the user to assume a fixed MBL a priori, or to allow the software to compute it locally from the geometry of the local computational cell, for example $\sqrt[3]{\text{cell volume}}$. This introduces a level of arbitrariness and mesh dependence. Some CFD packages avoid this ambiguity altogether by formulating the medium in terms weighted grey gases in a multigrey formulation (discussed below), and indeed, this preferred approach is becoming widespread.

Models of radiative properties may be broadly divided into three categories, listed in order of increasing computing effort.

a) Total or global models.
b) Spectral band models.
c) Spectral line-by-line (LBL) models.

Global or Total models include a variety of approaches. The simplest approach of all assumes a purely grey medium; despite its crudeness, it sometimes persists in yielding reasonable results provided the absorption coefficient is allowed to vary on the basis of a correlation of some sort. A more reliable approach is the multigrey or Weighted Sum of Grey Gases model (WSGG), originating with (Hottel and Sarofim, 1967) but developed further subsequently; this technique incurs much lower running costs than spectral models because the latter solve multiple RTEs.

Spectral band models include the exponential wide-band model (EWBM), narrow-band models (NBM), statistical narrow-band models (SNBM), and line-by-line (LBL).

LBL models involve lines measured in millions (with resolution of order 1 cm^{-1}), NBM (e.g. statistical narrow band and correlated-k) in hundreds (with resolution of order 25 cm^{-1}), wide-band models of O(10) with resolution of order 1000 cm^{-1}, and global models with several weighted gases.

Line-by-line methods are very expensive computationally, are not generally used in engineering simulations, and tend to be confined to atmospheric physics and climatology. Extensive databases are available in the public domain (Grosshandler, 1993; Soufiani and Taine, 1997; Clough *et al.*, 2005; Gordon *et al.*, 2017) and some focus on atmospheric applications. NBM are less expensive but are nonetheless very demanding computationally and are also relatively rare in engineering applications. Wide-band and global models are widely used in practical applications, especially the latter.

5.3.2 GLOBAL MODELS

Global models are the simplest and cheapest options available, and some incur errors which are low enough to be considered acceptable in real-world engineering applications, where the classic balance needs to be struck between accuracy on the one hand, and run time and costs on the other.

5.3.2.1 Grey Models

This is the crudest approach, with the RTE essentially being solved over one band – the infinite one. The absorption coefficient may, at its crudest, simply be a constant (the range of 0.3 to 0.5 m^{-1} is typical for hydrocarbon combustion at atmospheric pressure), or a linear function of temperature, e.g. (Hubbard and Tien, 1978), or higher-degree polynomials in temperature, e.g. (*Sandia Laboratory*, no date), or combinations of transcendental functions of temperature and soot concentrations (Gibbs and Joyner, 1978). There are also unpublished variants of the grey models.

The so-called integral or total model of radiative properties involves correlations derived by fitting the spectral data to obtain the total properties of each substance present in the mixture and then correcting for radiation overlaps, e.g. (Hottel and Egbert, 1942; Leckner, 1972; Felske and Tien, 1973; Modak, 1979; Steward and Kocaefe, 1986). Thus, for a mixture of water vapour, carbon dioxide and soot the emissivity is computed as $\varepsilon_{\text{mixture}}$ as follows:

$$\varepsilon_{\text{gas}} = \varepsilon_{H_2O} + \varepsilon_{CO_2} - \Delta_{\text{gas}} \tag{5.20}$$

where Δ_{gas} is a correction, and

$$\varepsilon_{\text{mixture}} = \varepsilon_{\text{gas}} + \varepsilon_s - \varepsilon_{\text{gas}}\varepsilon_s \tag{5.21}$$

The subscript s denotes soot. Early predictive methods, from around the middle of the 20th Century, focused on graphical charts for carbon dioxide and water vapour, named after Hottel who collated a large amount of experimental data obtained by himself and others (Hottel, 1954). Further analytical approximations (polynomials or transcendental functions) to the graphical charts have been offered, e.g. (Reeves, 1956; Lefebvre, 1984; Mehrota *et al.*, 1995), although some of these focus on gas turbine combustion chambers and higher temperatures. The subject of soot is elaborated on in Section 5.3.7. The issue of differences between emissivity and absorptivity of combustion mixtures, among other questions, is discussed in (Tam and Yuen, 2019).

5.3.2.2   Weighted Grey Gas Models (WSGG)

Quite good results have been obtained by expressing the emissivity (also known as emittance) as the WSGG (Taylor and Foster, 1974; Truelove, 1976; Docherty, 1982; Farag, 1982; Smith *et al.*, 1982; Coppalle and Vervisch, 1983; Modest, 1991; Rohsenow *et al.*, 1998), often abbreviated as WSGG or WGGM:

$$\varepsilon_g = \sum_{i=1}^{N} a_i(T)\left[1 - \exp\left(-K_i L_{MB}\right)\right] \tag{5.22}$$

N is the number of grey gases, which is typically 3 or 4. The parameters a_i and K_i have been derived by fitting this empiricism to either experimental data or the results of spectral models such as the wide-band and narrow-band models, or more accurate benchmarks, for absorbing-emitting media. A significant number of versions of this relation have been reported, for example with the weighting functions a_i being linear or quadratic functions of temperature, and N ranging from 2 to 4. The cited literature provides the parameters for a variety of species, and describes the procedure which would enable the current reader to develop Equation (5.22) for species which has not yet been reported. Here again, some of the references are confined to higher temperatures encountered in combustion chambers, and care must be taken to avoid applying the results outside the range of validity.

The grey WSGG derives an absorption coefficient from the above and solves one RTE. In the WSGG approach, sometimes referred to as multigrey or non-grey WSGG, several RTEs are solved as follows. The intensity associated with the ith grey gas is governed by (Modest, 1991):

$$\vec{\Omega} \cdot \nabla I_i + K_i I_i = a_i K_i n^2 I_B \tag{5.23}$$

Comparison with the RTE (Equation (4.41)) shows that this is just the transfer equation for a grey gas with an absorption coefficient K_i and with the blackbody intensity weighted by a_i. The parameters are independent of wavelength but may vary spatially and temporally during a CFD simulation. Equation (5.23) is solved for each of the grey gases, and the total intensity is the sum of the N grey intensities.

To summarise, the CFD modeller has two options:

a) Single-gas formulation: Solve for a single grey gas with absorption coefficient derived from Equations (5.19) and (5.22), after deciding how to specify the MBL, as discussed above. The literature also provides Planck-mean absorption coefficients (see Equation (4.1)):

$$K_{aP} = \frac{1}{I_B} \int_0^\infty K_{a\lambda} I_{B\lambda} d\lambda \tag{5.24}$$

Some workers have used Equation (5.19) directly, using emissivities derived from a spectral model, without resorting to Equation (5.20), e.g. (Orloff *et al.*, 1979). This is sometimes described as Grey Weighted Sum of Grey Gases (GWSGG).

b) Multigrey formulation: Solve the RTE for each of the N grey gases. This option is preferable because of its improved accuracy and avoidance of the MBL, albeit at a slight increase in central processing unit (CPU) costs.

The multigrey formulation of WSGG is reputable as a practical engineering tool, and is the formulation which is usually implied by the term WSGG (or WGGM) in the literature. It can be used with grey particulates and grey walls. It provides reasonable accuracy with a computational penalty which is not excessive. The WSGG has been improved upon in (Denison and Webb, 1993) and subsequent follow-ups. Many early models restricted the model ratio of H_2O to CO_2 to be either 2 or 1, associated with gas and oil combustion, respectively. Spectral models overcome that, but later developments of global models also allow arbitrary concentrations (Cassol *et al.*, 2014; Cassol *et al.*, 2015). Recent work (Guo *et al.*, 2021) confirms a preference for the non-grey WSGG rather than grey WSGG, and (Yang and Gopan, 2021) have used neural networks to improve on global models.

5.3.2.3   Full-Spectrum k-Distribution (FSK)
Several versions of this method, which is regarded as an improvement on WSGG, have been developed (Modest and Zhang, 2000, 2002; Modest and Riazzi, 2005).

The former is known as the Spectral-Line-Based WSGG (SLW), and the latter the FSK. The re-ordering discussed below in relation to k-d (k-Distribution) is performed over the whole spectrum rather than over spectral intervals, leading to a monotonic function logically called the full-spectrum k-distribution, e.g. (Ma *et al.*, 2014). Computational costs are comparable with those of WSGG, reducing the number of RT equations from millions to O(10), with accuracy sufficient and more amenable for engineering calculations, although derivation of the model parameters is more involved. An example of application of the method to produce a lookup table for a mixture of H_2O, CO_2 and CO is provided in (Wang *et al.*, 2016; Howell *et al.*, 2021).

5.3.3 BAND MODELS

Band models are generally more accurate than global models and also more expensive than the approaches already discussed, because the RTE is solved for each of the bands. Full details are beyond the scope of this book and can be found in the literature.

5.3.3.1 Wide Band Models

A variety of wide-band models (WBM) exist, as reported in the cited literature. The spectrum is divided into bands and the emissivity of each band is calculated once the gas composition, temperature, pressure, and mean beam length are specified. Note that the intrinsic parameters vary from band to band, and allow for any overlap. Computing effort is much smaller than narrow-band or LBL methods. The most successful version for combustion situations is the exponential wide-band model (EWBM), first developed in the 1970s, see (Edwards and Menard, 1964; Edwards and Balakrishnan, 1973; Edwards, 1976), and reviews in (Modest, 2021; Howell *et al.*, 2021). The number of bands is O(10), with a characteristic wavenumber separation of 1000 cm^{-1}. The closely packed rotational lines within the bands are re-ordered with respect to wavenumber, with their intensities decaying exponentially away from the band centre. The method includes the term 'wide' because the complete spectral range over the band is addressed. It exploits the observation that for combustion, infrared radiation is primarily due to up to six wide spectral regions for CO_2 and five for H_2O (see Table 5.1 above and the discussion preceding it).

The procedure begins with calculation of the band width of the absorbing/emitting species in the gas mixture, followed by ordering of the bands and accommodation of any overlaps, and finishing by computing the total mixture emissivity.

In the Grey-Wide-Band Model (GWB) of an absorbing-emitting medium the RTE is typically solved for between 6 and 9 bands (McGrattan et al., 2021), with an absorption coefficient for the *i*th band equal to the band Planck-mean value:

$$K_{ai} = \frac{1}{I_{bi}} \int_{\lambda_{i,min}}^{\lambda_{i,max}} K_{a\lambda} I_{b\lambda} d\lambda; I_{bi} = \int_{\lambda_{i,min}}^{\lambda_{i,max}} I_{b\lambda} d\lambda \qquad (5.25)$$

This is computed from a database such as those already cited (see Section 5.3.1). The intensity is then given by the algebraic sum of the band intensities.

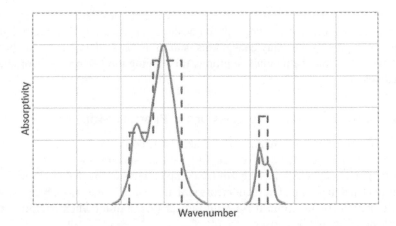

FIGURE 5.9 Schematic of several bands. Solid line is the narrow-band model, and the dashed line is the exponential wide-band model.

5.3.3.2   Narrow-Band Models (NBM)

Far more bands are involved in NBM, with their numbers measured in hundreds, and with spacing of the order of 25 cm^{-1}. A schematic illustrating the difference between WBM and NBM is provided in Figure 5.9. Spectral averaging is performed over the closely spaced vibration–rotation lines. Several NBMs exist, including Elsasser, Statistical (SNB), and k-distribution (k-d). Details are beyond the scope of this book and the reader can refer to the cited literature.

NBMs are more accurate than WBMs, and as expected, computing costs of NBMs are much larger than those of WBMs. SNB is accurate and is often used as a reference calculation against which the performance of other models are measured. Major reduction in computing time is reported by (Yan and Holmstedt, 1997), of the order of a factor of 20, for mixtures of CO_2, H_2O and soot and minimal loss of accuracy, by using lookup tables. That approach has been extended to include CO (Hofgren, 2015).

The k-distribution (KD) method provides band absorption coefficients directly. The values of the absorption coefficient within a typical spectral interval are repeated many times, and the method, originating in atmospheric physics in the 1930s (Fomin, 2004), involves re-ordering the line absorption coefficients in a narrow band into a smooth and monotonic function called the 'k-distribution', facilitating a quicker and cheaper integration over the interval using quadratures. The correlated version of KD (known as 'CKD' or 'Correlated-K' or 'C-K') is aimed at inhomogeneous situations and is therefore of less interest to CFD, which uses local values and tackles inhomogeneous flows computationally.

5.3.4   Line-by-Line Models (LBLM)

This is the most comprehensive and expensive approach, typically involving millions to hundreds of millions of lines and characteristic wavenumber separation of the order of fractions of a cm^{-1}. At this stage it is not considered to be a practical engineering tool for general and multidimensional engineering CFD, although the march of computer power may change this, and in any case the method can be used nowadays to generate benchmark solutions for judging the performance of cheaper alternatives.

5.3.5 SUMMARY OF PROS AND CONS OF THE PROPERTY MODELS

The first point to be made here, not for the first time in the literature, is that it is difficult to justify the CPU cost of a highly accurate spectral model used in a CFD simulation, when some other aspects of the complex physics involved in fluid flow, as well as heat and mass transfer, incur errors which are larger than those brought about by the radiation computation. It is useful to bear this in mind when deciding which spectral model should be chosen.

It should also be borne in mind that in a CFD simulation the RTE is solved for each band, and the spectroscopic properties need to be evaluated at every computational cell, given the temperature and chemical compositions. Some CFD packages create a radiation mesh which is coarse than the fluid mesh, typically made up of a cluster of the fluid cells, in order to reduce the CPU cost. The spectroscopic properties need to be computed at every iteration cycle, and every time step in a transient simulation. LBL is not a practical option for the usual CFD scenarios, and the same applies to most NBMs, although the speed-ups already cited above could change this view. LBL and NB models can be used to generate benchmark solution for testing alternatives.

The FSK model is popular as a practical tool with a good balance between accuracy and cost. Another choice at that level of compromise is the Weighted Sum of Grey Gases (WSGG or WGGM). The single-band grey version called the Grey Weighted Sum of Grey Gases (GWSGG) is cheaper to run and less accurate than the above.

Useful comparisons of accuracy and computing time for several test cases are provided by (Demarco-Bull, 2012) where conclusions by other workers are generally confirmed and elaborated on. The various spectroscopic models, in grey, wide-band and global categories are compared with a reference SNB setup involving 367 bands, for a flame. The quoted 'CPU ratio' is the CPU time divided by the reference (SNB) CPU time. The grey models yielded maximum error between 32% and 61%, mean errors between 8% and 17%, and CPU ratios between 1/58 and 1/216. The figures for the global models were 6% to 14%, 1.3% to 1.8%, and 1/20 to 1/52, respectively. Of the global models, WSGG yielded 6.25%, 1.81% and 1/52, respectively, which is attractive if a compromise is sought.

5.3.6 POROUS MEDIA

The literature on porous media in general is substantial, and applications are wide-ranging. Just a few general references will be cited here (Ingham and Pop, 1998; Vadász, 2008; Kaviany, 2012; Scheidegger, 2020) because this book focuses on thermal radiation, with a very brief outline of the general aspects.

Usually, the term 'porous medium' refers to one made up of solid and fluid regions, although sometimes the porous concepts are applied to fluid–fluid (e.g. liquid–gas systems); examples of applications are:

- Thermal insulation (e.g. foam and fibrous).
- Packed/fluidised bed reactors.
- Foliage, e.g. crops and forest canopies.
- Tube banks when modelled on a sub-grid basis as a continuum, e.g. heat exchangers and steam generators.
- Geothermal power.
- Biofuel combustion.
- Drying of porous materials in the food and textile industries.
- Soil remediation and air sparging.
- Degraded nuclear fuel.

Radiation is not necessarily important in all of these examples.

The volume porosity is defined as the ratio of the fluid volume to the total volume, and may be defined on a global or local value, with the latter applying when the computational cells are much larger than the characteristic size of the pores. A system with a high porosity (close to 1.0) may involve situations referred to as 'dispersed', 'dilute' or 'particle-laden' in the various sectors and industries.

The first important concept which should arguably be highlighted here is the distinction between continuum and heterogeneous media. In the former, the porous spatial scales are small relative to the size of the computational cells. Hydrodynamically, this is related to the ratio of the pore size to the overall dimension of the domain being considered. However, when thermal radiation is added to the problem, one must also consider the ratio of the pore size to the characteristic electromagnetic wavelength. If that ratio is large, then the medium is heterogeneous from the radiation viewpoint. Moreover, the solid's morphology will dictate whether electromagnetic scattering will be dependent or independent (Tien and Drolen, 1987).

When the pore size is much smaller than the computational cell size, the analysis is a continuum one. However, it requires the spectroscopic properties of the continuum. Typically, such a medium consists of a porous solid, with the pores being filled with a fluid which is transparent on the scale of the pores. Such a medium is sometimes described in the literature as 'semi-transparent' (cf. Section 4.6). In some cases the solid is not opaque, and the fluid in the pores is not transparent.

Another important concept in the field of porous media distinguishes between equilibrium and non-equilibrium phenomena in such scenarios. Focusing now on situations involving flow of single or multiphase fluids through a porous solid, in an equilibrium situation the only entity which is exchanged between the fluid and the solid is momentum. An example is isothermal flow of water and air though soil and bedrock. More complexity is introduced if the solid exchanges heat or mass with the fluid. Thus, if the solid's temperature differs from that of the fluid, the situation lies in the non-equilibrium regime, and generally, a theoretical analysis must account for heat and/or mass transfer in the solid as well as interfacial exchange. There are limits of this regime which can still be described as equilibrium:- For example, clouds of dilute particulates can be regarded as a dilute limit of a porous medium, and if the particulates are sufficiently small then the heat exchange with the fluid in which they are carried is very efficient and the particulates' temperature is very close to that of the fluid.

Turning to some of the sample applications in the bulleted list above, non-equilibrium is illustrated in the case of a tube bank, say in a heat exchanger or boiler, which can be analysed in several ways. A continuum approach models the shell side as a porous medium, coupled thermally to a separate model, usually an idealised analytical or semi-analytical one, of flow and heat transfer on the tube side, possibly with a 1-D or 2-D model of conduction in the tube walls too. Given the rapid advances in the power of both hardware and software, it is not uncommon nowadays for a non-continuum approach to be used, with explicit modelling of both the flow and heat transfer on the shell and tube sides (sometimes with multiphase effects such as boiling), as well as conduction in the tube walls. The simultaneous and automatically coupled analysis of heat transfer in fluids and adjacent solids is known as 'Conjugate Heat Transfer' (CHT). Naturally, if the mass flow on the tube side (or indeed the shell side) is very high, or if say the fluid possesses high thermal inertia and conductivity (such as liquid metal), then the interface temperature may be essentially uniform, which simplifies the analysis.

The simplest engineering approach to the situation commonly described as 'a fluid saturated porous medium' regards the medium as a continuum (already introduced above), made up of a conducting solid which is intimately coupled to the fluid. The radiative field in the medium usually involves absorption, emission, and scattering. The opacity of the fluid in the pores may or may not be negligible, but the classical work deals with small pore/fluid opacity. Pioneering early papers (Larkin and Churchill, 1959; Chen and Churchill, 1963), for the classical situation, developed a model which generalised earlier work and compared theory with experimental measurements of packed beds (made up of spherical particles of various materials). As far as thermal radiation was concerned, the model led solely and conveniently to an enhancement of the thermal conductivity of the continuum by a 'radiation conductivity' k_R, of the form

$$k_R = 4F\sigma DT_S^3; F = \frac{2}{(a+2b)D} \tag{5.26}$$

where T_S is the local source (solid) temperature, and D is the particle diameter. The parameters a and b are the radiative absorption and back-scattering cross-sections per unit volume of packing respectively (m^{-1}). In earlier work F was predicted to be either a constant (e.g. 1/3) or only a function of the particles' surface emissivity, whereas the cited work showed that F is affected by the particle shape, size, and transmissivity.

The above work has been extended by others; see for example a brief review in the online Appendix E in (Howell et al., 2021), and (Singh and Kaviany, 1992; Travkin and Catton, 1999; Zeghondy et al., 2006; Lipiński et al., 2010; Sacadura, 2011; Zhang et al., 2014; Park, 2016). The work of (Lee, 1989; Marschall and Milos, 1997;

Lee and Cunnington, 2000; Linder, 2014; Dauvois *et al.*, 2017) are examples of fibrous media applications, and (Glicksman and Torpey, 1988; Baillis *et al.*, 2000) of foams. The Monte Carlo method has often been used to derive the continuum's spectroscopic properties by analysing a small region of the domain in full, heterogeneous detail.

Radiation in dispersed media, assumed here to involve independent scattering, has received much interest, e.g. (Viskanta and Menguc, 1989; Zenier *et al.*, 2001; Dombrovsky and Baillis, 2010; Al-Gebory and Menguc, 2020).

5.3.7 Particulates

The term 'particulates' is being used here in a general sense, not necessarily spherical, and may for example be soot, dust or water droplets.

The radiative behaviour of individual particulates depends on the material's properties and on the ratio of the particle size to the electromagnetic wavelength; whilst generally this results in absorption, emission, and scattering, the latter varies like the inverse of fourth power of wavelength when the above ratio is small and can be ignored in some cases for small particles on these scales.

The classical theory of the interaction between a spherical particle and an electromagnetic monochromatic wave is named after Mie, even though it was predated by the work of Lorenz, e.g. (van de Hulst, 1981). The reader is referred to the material in section 4.1.10 on scattering, and the Rayleigh, Lorenz-Mie-Debye, and geometric optics regimes. Equation (4.34) is repeated here for convenience:

$$\hat{d} = \frac{\pi d}{\lambda} \tag{5.27}$$

The interaction of an incident electromagnetic wave with a cloud of particulates involves the following contributions to scattering:

- Reflection from the particles.
- Refraction of waves passing through the particle.
- Diffraction of the wave due to the proximity of the particles to one another.

If the separation between the particles (when expressed as the ratio of the separation to the wavelength, denoted by $\hat{L}_P$ in Equation 4.35, and to the particle size) are small or large, the scattering is said to be dependent and independent respectively. The theory for a single particle, involving $\hat{d}$, is extended to a cloud of particles to derive the radiative property of the fluid–particle mixture. This usually assumes that the cloud is dilute. Non-dilute theories have however been developed for a variety of situations such as porous media discussed above (e.g. fibrous material and packed beds) and individual soot clusters, e.g. (Ma *et al.*, 1990), and the discussions on porous media and soot elsewhere in this book.

As an example of a dilute system, consider a dilute cloud (also named 'polydispersion') made up of spherical particles with a size distribution $N(r)$, where r is the particle radius, and $N(r)dr$ is the number of particles per unit volume which lie in the size range r to $r + dr$. The cloud extinction coefficient K_p is then given by

$$K_p = \pi \int_0^\infty r^2 Q(r) N(r) dr \qquad (5.28)$$

where Q is the single-particle extinction efficiency (see Section 4.2). Analogous expressions apply for the absorption and scattering coefficients. In typical engineering simulations the cloud absorption and scattering coefficients are added algebraically to those of the continuum in which the particles are suspended.

There are many examples involving liquid drops or droplets, and only a few will be cited here. In the discipline of climate change, the role of clouds is important of course, e.g. (McKee and Cox, 1974; Welch *et al.*, 1980; Arking, 1991; Wielicki *et al.*, 1995; Randall *et al.*, 2007; Davis and Marshak, 2010; Hogan and Shonk, 2013; Okata, 2018). The term 'cloud' can refer to liquid droplets or ice crystals typical of cirrus clouds. Other particulates such as dust and soot are also relevant of course. In fire suppression by water spray, the water droplets attenuate the radiation, but also evaporate to increase the water vapour concentration; some analyses also include the effects of hydrodynamic interactions of the water droplets with the soot particles (Sinai *et al.*, 2008).

Soot merits special mention because of its importance in industrial combustion, fire and climatology. The relevance of soot will diminish somewhat as the use of fossil fuel declines, but increasing occurrence of natural fires is being predicted, and combustion of various fuels will continue. As an example, a view of the 'Magnum' forest fire provided in Figure 5.10. The luminescence at the base of the plume is caused principally by soot within the flame itself, which increases the radiative emission from the flame, and the thick black smoke at higher levels is associated with soot which has grown to larger sizes (due to agglomeration) and enshrouds the flame in the colder regions as the plume mixes with the ambient atmosphere.

Soot particles are made up of very small primary carbon particles (also called monomers), near-spherical in shape, measured in nanometres or tens of nanometres. Generally, however, these particle agglomerate to form chain-like clusters with the cluster size usually being reported to be up to several microns (Mulholland, 1995; Shaddix and Williams, 2007; Snelling *et al.*, 2011; Altenhoff *et al.*, 2020); some research has reported clusters measuring hundreds of microns (Sorensen and Feke, 1996), but questions have been raised in the literature about the sampling methods used in the measurements. Be that as it may, this all takes the spectroscopic behaviour beyond the range of the Rayleigh approximation, and also into the dependent scattering regime. The phenomenon is complicated by condensation and/or evaporation of water and other species on the particles, reaction with trace gases, or agglomeration with other clusters, and images are provided in Figures 5.11 and 5.12 (Hu and Köylü, 2004; Radney *et al.*, 2014). The morphology of the two clusters in Figure 5.12 are described as 'lacey' and 'compacted', respectively. The research has involved fractal morphology, e.g. (Jullien and Botet, 1987),

FIGURE 5.10 'Magnum fire', June 2020, at Kaibab National Forest, Arizona, USA. The flame luminosity is unimpeded at the base, but attenuated by relatively cold soot enshrouding the flame at higher regions. Reproduced from Flickr.

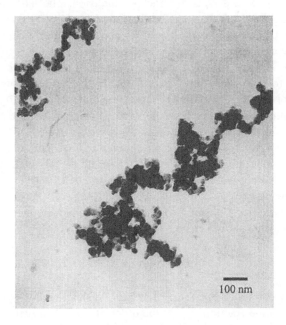

100 nm

FIGURE 5.11 Image of soot agglomerates generated by an acetylene flame (Hu and Köylü, 2004). Reproduced with permission of the American Association for Aerosol Research.

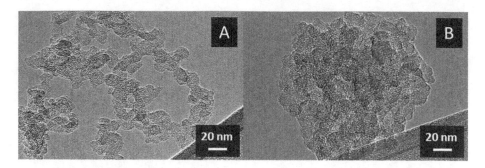

FIGURE 5.12   Images of soot agglomerates. Left: 'Lacey' morphology. Right: 'Compacted' morphology. Reprinted with permission from (Radney *et al.*, 2014). Copyright (2014) American Chemical Society.

leading to relationships for the number of primary monomers in each cluster depending on the fractal dimension, e.g. (Radney *et al.*, 2014):

$$N_{\mathrm{mon}} = k_0 \left(\frac{2R_G}{D_{\mathrm{mon}}} \right)^{D_f} \tag{5.29}$$

where k_0 is a fractal prefactor, R_G is the radius of gyration, D_{mon} is the diameter of the primary monomer, and D_f is the fractal dimension. The latter is 1 for a linear agglomerate and 3 for a spherical agglomerate, and a value of 1.8 is typical for soot.

Soot can affect the spectroscopic properties of a medium dramatically, by increasing the effective extinction coefficient. In combustion, that leads to a change from non-luminous to luminous flames, and emission from the latter is generally much larger than the former. Early theoretical work on the radiative properties of soot ignored the major complexity of the agglomeration and non-sphericity, and employed the Rayleigh limit (diameter much smaller than wavelength) of the Mie theory (van de Hulst, 1981; Van de Hulst, 2012) for a single carbon sphere, which involves the real and imaginary parts of the complex refractive index (Section 4.1.3). These two parameters were measured by (Foster and Howarth, 1968; Dalzell and Sarofim, 1969). The latter's data were used by several workers to derive Planck-mean and Rosseland-mean absorption coefficients for soot cloud, e.g. (Felske and Tien, 1973; Hubbard and Tien, 1978; Tien and Lee, 1982), leading to a simple effective grey soot absorption coefficient (in m^{-1}) given by

$$K_S = 1270 f_v T \tag{5.30}$$

where f_v is the local soot volume fraction and T is the local mixture temperature (Kelvin). Soot volume fractions are of the order of 10^{-6}. The literature reports soot mass-specific extinction cross-sections measured approximately in the range 5–15 m^2/g. The extinction coefficient is obtained by multiplying this value by the soot concentration in g/m³. The modelling approaches which are simpler than WSGG simply add the value in Equation (5.30) algebraically to the extinction coefficient of the gaseous components. The reader is referred to Sections 5.3.2.1 and 5.3.2.2 for related material.

Over the past three decades workers have addressed the challenging spectroscopic behaviour of the agglomerates, and this activity is ongoing, e.g. (Ku and Shim, 1991; Faeth and Köylü, 1995; Farias et al., 1995; Krishnan et al., 2000; Eymet et al., 2002; Liu et al., 2008; Maugendre et al., 2008; Chen et al., 2010; Radney et al., 2014; Suryadharma, 2020). Interestingly, some of the techniques originated in the field of crystallography, aspects of which are related to issues arising in porous media discussed above. The literature reports scattering predictive techniques known as the Rayleigh–Debye–Gans fractal-aggregate theory, the T-matrix Method, Generalised Multi-sphere Mie, and Discrete Dipole Approximation, all of which are beyond the scope of this book.

5.4 SOLUTION TECHNIQUES

As with spectroscopic properties, a distinction is logically made between transparent and participating media. In fact, as already discussed, an intermediate category of substances and situations exists, usually referred to as semi-transparent, such as glass.

The RTE covers the complete range and is the most general formulation. Participating situations are much more complex mathematically than transparent ones, especially but not only when scattering is present. A variety of techniques have been developed over the years for both extremes.

Since transparent situations are relatively common and simpler to solve, limited mathematical detail is provided here, but even for participating scenarios only a very brief outline is provided, and the interested reader is referred to the literature on thermal radiation which has already been cited, as well as (Raithby and Chui, 1990; Rohsenow et al., 1998; Tencer and Howell, 2016).

Table 5.2 provides a list of common solutions methods with some comments. The focus is on methods which can tackle three-dimensional problems. The comments are aimed at users of radiation solvers which are already implemented in the CFD suite which is being considered, rather than at developers aiming to implement a radiation solver in a CFD package. Capabilities of the methods apply in the general sense, and the user of course needs to check what level of capability has been implemented in the software being considered or used.

A pitfall of some methods is the so-called 'ray effect'. This is an unrealistic phenomenon which leads to highly inhomogeneous intensity fields due to directional discretisations, especially when the situation involves a highly localised source region which is much smaller than the domain as a whole. This matter is discussed in more details in Section 5.4.2.11.1.

5.4.1 TRANSPARENT MEDIA

A few basic concepts relating to the electrical analogy were introduced in Section 4.1.5, where it is also noted that the method was first put forward by (Oppenheim,

1956) and has been extended and generalised. A little more detail is provided in Appendix B. The method is still being used in engineering calculations but numerical radiation solvers (discussed below) which are often included in CFD packages are replacing this technique.

Common methods for transparent media are view-factor and electrical analogy, Discrete Transfer, and Monte Carlo. 'View factors' are sometime referred to as 'shape factors', 'angle factors', or 'configuration factors'. When all the surfaces are black (perfectly absorbing), these factors are 'geometric' factors and only involve the geometry of the domain. A common notation is F_{i-j}, which means the fraction of the radiant energy leaving surface 'i' which reaches surface 'j'.

The electrical analogy approach enables what seems to be a formal solution to the problem, but it involves view factors, and for other than the tabulated geometries, whether in analytical or graphical form, numerical methods are required to evaluate the view factors, so the electrical analogy in itself is not a solution to the RTE. Electrical analogy methods have been demonstrated for black diffuse surfaces, grey diffuse surfaces, combinations of grey diffuse and specular surfaces, and even for a participating medium inside the domain. The analogy becomes increasingly involved as more of these various physical effects are included, and it is becoming harder to justify such methodology when numerical solvers are commonly available, often able to account for a wide variety of physics beyond the scope of the electrical analogy or net radiation method. For this reason, only a brief outline has been provided in this book.

The 'S2S' model is essentially the network model, with 'clustering' of radiating boundaries in order to reduce the high computational cost if, say, every boundary cell face were defined as a surface. The cluster temperature is calculated as an area-weighted average of the individual face temperatures raised to the fourth power. The model then evaluates view factors using Equation (4.13), and surface radiative fluxes are computed as in the network model.

5.4.2 Participating Media

As explained above, because radiative transfer in participating media tends to be much more complex than in transparent media, this section is generally brief. For deeper (and relatively modern) information, the reader is referred to the references cited in Section 4.1, and especially (Modest, 2021; Howell *et al.*, 2021).

It should be remembered that the radiation solver needs to provide two entities to the flow solver: Radiative fluxes at boundaries, and the divergence of the radiative flux inside the domain (see Section 4.4).

RTE solution techniques may be divided into two broad categories:

• Deterministic methods.
• Stochastic (statistical method).

Referring to Table 5.2, numbers 1 to 9 fall into the first category, and Monte Carlo into the second.

TABLE 5.2
Some Radiation Solution Techniques

No.	Name	Acronym	Size of Opacity
1	Network	—	Zero/small
2	Surface-to-Surface	S2S	Zero/small
3	Rosseland	—	High
4	Six-flux	—	Any
5	Zonal	—	Any
6	Finite Volume	FV	Any
7	Spherical Harmonics	P_N	Any (but generally poor at low opacities)
8	Discrete Ordinate	DO, S_N	Any
9	Discrete Transfer	DT	Any
10	Monte Carlo	MC	Any

To the author's knowledge the early solution techniques for participating media were developed in the field of astrophysics (3 and 4 in Table 5.2), and tended to focus on optically thick (diffusive) situations.

5.4.2.1 Rosseland

The Rosseland model does not solve a separate radiation equation, and instead represents the radiative effects by simply enhancing the medium's thermal conductivity by a radiative conductivity. The latter is proportional to the local temperature raised to the power 3, and is inversely proportional to the extinction coefficient. Thus, the total energy flux in the domain, away from boundaries, is simply given by

$$q = -\left(k + k_R\right)\nabla T; k_R = \frac{16\sigma n^2 T^3}{3K} \qquad (5.31)$$

The Deissler jump conditions (Deissler, 1964) improves on the large errors at boundaries.

5.4.2.2 Schuster–Schwatzchild

Moving on to general opacities, in the order in which the techniques are listed in the above table, the six-flux method is a generalisation of the Schuster–Schwarzschild technique, which is a forerunner of the Discrete Ordinate method. Schuster's and Schwarzschild's results, obtained independently, addressed non-scattering media, although this has been generalised to isotropically scattering media (Modest, 2021). The analysis is one-dimensional for the plane-parallel slab, with directions discretised in only two directions, entailing two hemispheres, which is why the technique is also known as the two-flux method.

Angular resolution is poor in flux methods (unless scattering is strong), encouraging users to focus on specific geometries (e.g. cuboids), and the technique is therefore of limited use in general geometries.

5.4.2.3   Zonal

Hotel's zonal method relies on spatial discretisation, unlike most methods which employ discretisation of direction. In general the exchange factors, between volume elements and boundaries and between the volume elements themselves, need to be repeatedly computed. It was a leader in its time but has now been superseded.

5.4.2.4   Finite Volume (FV)

The Finite Volume Method (FVM) possesses a natural connection with CFD, as its name implies, and is a viable method. At each cell the RTE is integrated over the cell volume and direction, leading to a surface integral over the cell faces. Intensities on the cell faces are related to those at neighbouring cells, which introduces numerical diffusion which renders the method less vulnerable to ray effects than the ray tracing approaches.

5.4.2.5   Spherical Harmonics (P_N)

In the P_N approach (also known as a Moment method) the intensity is written as a rapidly converging series of orthogonal spherical harmonics (Abramowitz and Stegun, 1965), and the method produces smooth solutions. A small number of terms usually suffices, and even P_1 (also known as the differential approximation) in which the zeroeth moment and first order moments are deployed, is offered in most CFD packages. This leads to a partial differential equation for the spectral or grey incident radiation. A caveat is that if major errors are to be avoided in P_1 it is essential to use the so-called radiative slip or jump boundary conditions, in which there is a discontinuity of the dependent variable at the boundaries (Deissler, 1964). In general the P_1 errors can be large, although the model does much better in optically thick situations. The paper by (Ravishankar et al., 2010) reports that overall, P_3 is more accurate than P_1, although for test cases in which the opacity was low (0.1) both models fared poorly.

As an example, the governing equation for a non-scattering medium, in terms of the incident radiation G (at any interior point) and in the context of the P_1 approximation, is quoted here:

$$\frac{1}{K^2}\nabla^2 G - 3G = -12\pi I_b \tag{5.32}$$

This is an elliptic equation for the variable G, which is compatible with typical CFD solvers. A generalisation to a linearly anisotropic scattering medium exists, e.g. (Modest, 2021).

The so-called Simplified Spherical Harmonics Method (SP_N) was first outlined in Westinghouse nuclear reports in the 1960s, and involved an ad-hoc change of the Spherical Harmonics expansion for the plane slab to a 3-D form. It has been used increasingly in nuclear engineering, e.g. (Brantley and Larsen, 2000; Chao, 2016), although investigation thereof has entered the heat transfer sector too (Modest et al., 2014). SP_1 is identical to P_1, and the differences manifest from SP_3 and on to higher moments.

5.4.2.6   Discrete Ordinates (S_N)

As alluded to above, in the Discrete Ordinates methods (S_N) the direction is discretised, and the term 'ordinate' refers to direction. The scattering integral is approximated in terms of quadratures and weighted sums; thus

$$\text{Scattering term} = \frac{K_s}{4\pi} \sum_{i=1}^{n} \phi_i I \tilde{p} \tag{5.33}$$

where n is the number of directions and ϕ_i are the weights. For each ordinate there is a second ordinate in the opposite direction. A transport equation is solved for each direction $\hat{s}_i$. Referring to Equation (4.41), this is of the form

$$\hat{s}_i \cdot \nabla I + KI = K_a n^2 I_b + \frac{K_s}{4\pi} \sum_{i=1}^{n} \phi_i I \tilde{p} \tag{5.34}$$

where the intensity I and the scattering phase function $\tilde{p}$ are both functions of location and the direction. In 3-D, each octant is discretised in the azimuthal and polar directions, and the number of RTEs which are solved equals 8 times the number of ordinates in an octant. Thus, with 2×2 discretisation 32 RTE are computed, and with 10×10 discretisation the number is 800, and computing effort increases accordingly. Quadratures are employed in the boundary conditions too. That set, together with the set in Equation (5.34), make up a coupled set of n simultaneous, linear partial differential equations for the intensity as a function of distance along a direction, for each of the n directions. Coupling due to scattering renders the numerical solution more complex. Remember too that two-way coupling exists between the intensities and the temperature field. Once the intensities are determined the radiative fluxes, whether volumetric (i.e. inside the domain) or at boundaries, are derived through quadratures.

5.4.2.7   Discrete Transfer (DT)

The discrete transfer method (DTM) can be described as a hybrid method. Rays are traced from surface elements, through volume elements towards other boundaries, and interact with the volume elements. Essentially, Equation (I.3), generalised to a spectral form if necessary, is applied across a cell as a recurrence relation. The emission and absorption, and isotropic scattering, in a given computational cell is then obtained from the values thereof for all the rays which traverse that cell, and the process repeated iteratively.

5.4.2.8   Monte Carlo (MC)

The Monte Carlo (MC) method is a stochastic one, in which random bundles of photons are tracked through the system. Bundles are emitted from surfaces and emitting volume elements, and tracked until they are absorbed or leave a relevant boundary (such as a transparent or semi-transparent entity, or an opening). Probability distribution functions, together with random numbers, are used to choose a full variety of physical attributes pertaining to the bundle birth and its fate throughout its life-time, such as source location, wavelength, absorption, scattering, scattering angle, and behaviour at boundaries. The method can thus deal with any level of physics complexity, including spectral effects, anisotropic scattering, and collimated and diffuse radiation. Unsurprisingly, there is a potentially heavy price to pay compu-tationally, although computing power growth is such that use of MC in practical sce-narios is increasing rapidly. Research continues, as illustrated by (Torres-Monclard *et al.*, 2021), who report on statistical enhancements which reduce CPU, and include fractal-aggregate theory to model the non-spherical morphology of soot particles, as outlined in Section 5.3.7.

5.4.2.9 Hybrid Methods

The poor performance of the popular P_1 model when the opacity is low has already been highlighted. Moreover, the ray effects, already mentioned above, and elaborated on in Section 5.4.2.11.1, have been a drawback. These issues have prompted attempts to overcome the weaknesses of one technique with the strengths of another. This has led to a class of methods known as 'hybrid' or 'combined'. The former term is used rather widely, and DT has been described in the literature as a hybrid method which exploits the benefits of virtues of MC, DO and zonal methods, but this section con-centrates on hybrid methods specifically designed to perform well across the whole range of opacity, and possibly to minimise the ray effects too.

A description of early attempts at hybrid methods may be found in (Viskanta and Menguc, 1989), and a wider review accounting for more recent developments is available in (Modest, 2021). The latter describes a seminal and promising approach which was first enunciated by Olfe and co-workers (Olfe, 1967; Glatt and Olfe, 1973), and was described as a 'modified differential approximation', abbreviated as MDA. That initial work dealt with grey non-scattering media within grey enclosures, which was later generalised, but in any case unless scattering is highly anisotropic, intensity originating from walls varies significantly with directions, whereas inten-sity emitted and scattered by the volumetric source tends to vary more slowly with respect to direction. For that reason, the intensity was decomposed as the algebraic sum of two fields:

$$I = I_w + I_m \tag{5.35}$$

Here the first term is a field associated with emissions from walls, which may be attenuated by reflections from other parts of the walls or by absorption within the medium. The second term is a field associated with the medium and associated volu-metric radiative source, i.e. emission. The wall field along a ray is given by Equation (I.1) in Appendix I, and when this is translated to radiosity in order to solve for the wall field in a general 3-D environment using the electrical analogy, one is led to the classical set of linear simultaneous equations for the radiosities (Equation (B.1) in Appendix B) except that the view factor in the summation term is weighted by the attenuation term $\exp(-\tau_{sij})$, where τ_{sij} is the optical depth between surface i and surface j. The medium's field was evaluated using the P_1 method. Thus, the wall field required the network model modified by the exponential, and the local incident wall and medium fields supplied G in Equation (5.32). The technique led to several benefits: It behaved well across the whole spectrum of opacities, and avoided the ray effects.

Isotropic and linear-anisotropic scattering, and generalised wall behaviour, was subsequently added by other workers (Wu et al., 1987; Modest, 1989). Scattering adds significant complexity. The literature has reported high accuracy by the method, e.g. (Park et al., 1993). Some papers in the literature, for example (Dombrovsky, 2002), refer to a modified differential approximation but the methodology is a flux-based method and is unrelated to the techniques being discussed here. The MDA approach has been enhanced further by the so-called Improved Differential Approximation (IDA), see (Modest, 1990).

Other workers have explored other combinations of methods besides radiosity + P_1. For example, in (Baek et al, 2000), the medium's field is solved by the Finite Volume Method, and the wall-related field is computed using MC instead of the radiosity/electrical analogy method.

5.4.2.10 Scattering Capabilities

The diffusion model and the zonal model can only handle isotropic scattering, and P_1 can handle linear anisotropy too. All the other methods can cope with anisotropic scattering, the complexity of which depends on the accuracy of the directional scheme of the method.

5.4.2.11 Examples of Some Pitfalls

The Monte Carlo method is considered to be the most comprehensive RTE solver. However, it tends to be expensive when a large number of bundles are used, as is required for high accuracy. Some pitfalls of the remaining methods will be outlined in this section.

5.4.2.11.1 Ray Effects

Any method involving discretisation of direction will, in principle, introduce numerical behaviour which reflects that discretisation. This is known as the ray effect, e.g. (Chai et al., 1993; Seung and Kang, 1999; Coelho, 2001; Tencer, 2014; Hunter and Guo, 2015; Kamdem, 2015).

For example, in a turbojet combustion chambers hot gases fill all or most of the chamber, and the heat fluxes at the chamber walls are unlikely to be strongly affected by ray effects. On the other hand, a gas flare burning in the atmosphere involves radiation propagating from the flare to remote locations, where information thereon is required in order to assess heat loading on structures or personnel. At those remote locations the delivered radiation predicted by DT and DO will possess an inhomogeneous pattern, sometimes star-shaped. This is helped by increasing the number of angular discretisations or refining the mesh, but the computing cost may rise dramatically, especially in the case of DO. MC is also susceptible to a 'blotchy' pattern if the number of photons or bundles/histories is not high enough; the typical recommended number of bundles for engineering simulations is in the millions to hundreds of millions, although this must be judged on a case-by-case basis.

Figures 5.13 and 5.14 illustrate these effects. They portray the irradiation at the ground below a low-momentum methane flare at a chimney stack, subject to a cross wind. The ray effect is particularly conspicuous when the source region is close to a boundary, and when the medium is non-scattering. Scattering alleviates the ray effect.

For comparison, the results from Monte Carlo simulations of the same problem are shown in Figures 5.15 and 5.16.

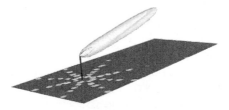

FIGURE 5.13   Irradiation at the ground, discrete transfer, 8 × 8 rays.

FIGURE 5.14   Irradiation at the ground, discrete transfer, 24 × 24 rays.

FIGURE 5.15   Irradiation at the ground, Monte Carlo, 10^6 bundles.

FIGURE 5.16   Irradiation at the ground, Monte Carlo, 10^7 bundles.

5.4.2.11.2   False Scattering

This phenomenon is analogous to the familiar false diffusion in CFD, whereby discretisation errors manifest as diffusion which, if not treated with care, can dominate real diffusion. Mesh refinement is required, with accompanying costs of course.

5.4.2.11.3   Other Issues

In an elongated geometry the rays traced from a narrow part of the boundary capture large areas of the longer part of the geometry, unless the hemisphere at boundaries is not discretised uniformly. This leads to lower resolution and accuracy at some locations of the domain and is related to the ray effect.

In geometries which are very complex, numerical difficulties are sometimes experienced at corners. It is recommended that iso-surfaces of temperature and radiation intensity are visualised, which is in any case something that should be done as part of standard quality assurance (QA) checking; this will reveal any unexpected behaviour which may be hidden somewhere in the depths of the domain. If such challenges persist, consider deployment of a hybrid method or even a diffusive solver such as P_1 or P_3, without losing sight of the different types of errors which are likely to be introduced by doing so.

5.4.3   COMPARISONS OF THE RTE SOLUTION METHODS

Many articles have been published on comparisons of the various RTE solution techniques, and the reader is referred to the literature already cited and to (Singh and Mishra, 2007; Joseph *et al.*, 2009; Coelho, 2014).

Table 5.3 provides a list of the principal RTE solvers in use nowadays, together with their strengths and weaknesses. This information aims to aid the practitioner in choosing an appropriate solver, without losing sight, of course, of the spectroscopic properties model which also needs to be chosen, as discussed in Section 5.3.

The following strategy is therefore suggested, with the aim of compromising between accuracy and computing cost and/or computing time if project time scales and budgets do not permit usage of the more accurate options available to the practitioner.

If the emitting regions occupy a small part of the domain:

a) If radiation at locations far from the emitting regions (such as remote boundaries) is important then increase the number of directional discretisations when using ray tracing methods, or number of bundles when using MC. This will increase the cost and run time of the radiative component of the simulation. For example, the default number of angular discretisations of the hemisphere for DT and DO is of the order of 64 (8 × 8) in commercial packages; this may need to be increased to about 2500 (50 × 50). The number of MC bundles may need to be hundreds of millions say, or more. If the cost and run time impact of such resolution is impractical, then avoid DO and DT, and consider alternatives such as P_N, whilst bearing in mind their deficiencies. An exception might arise when the boundary temperature is fixed and prescribed, or if the boundary is the interface between the fluid

TABLE 5.3
Principal Radiation Solution Techniques, and Their Strengths and Weaknesses

No.	Name	Acronym	Size of Opacity	Compatibility with Spectral Models. Scattering	Comments
1	Network		Zero/small	—	'Reformulation' rather than solution: requires view factors, tabulated or pre-computed.
2	Surface-to-Surface	S2S	Zero/small	—	Computing costs reduced by 'clustering'.
3	Rosseland		High	General spectroscopy. Isotropic.	Diffusion regime. Cheap, & accurate at high opacities, except near boundaries. Singularity at spectral windows.
4	Six-flux		Any	Most spectral models. Isotropic.	Generalisation of the two-flux Schuster–Schwarzschild method. Mainly cuboidal geometries. Forerunner to DO. Lower accuracy.
5	Zonal		Any	Grey, or WSGG. Isotropic.	First general method. Becoming outdated.
6	Finite Volume	FV	Any	Most spectral models. Any scattering.	Good synthesis with CFD.
7	Spherical Harmonics	P_N	Any (but P_1 is generally poor at low opacities)	Most spectral models. Linear anisotropy.	P_1 is also known as the 'differential approximation'. Smooth solutions, cheap.
8	Discrete Ordinates	DO, S_N	Any	Most spectral models. Any scattering.	One of leading methods. Suffers from ray effects.
9	Discrete Transfer	DT	Any	Most spectral models. Isotropic.	One of leading methods, suffers from ray effects.
10	Combined/Hybrid	MDA, IDA, plus others	Any	Grey, but spectral possibilities. Linear anisotropy.	Promising approach, avoids ray effects.
11	Monte Carlo	MC	Any	Any spectral models. Any scattering.	Statistical technique. One of leading methods. More expensive.

domain and a CHT object; conduction in the latter may 'smear' the non-uniformities caused by the ray effects in the fluid. It is difficult to generalise here and the practitioner needs to assess these issues in each case.

b) If radiation at locations far from the emitting regions are not important, then consider DT and DO. One example is an unconfined or semi-confined fire, when the key deliverables are conditions within and very close to the flame and smoke products (for example heat delivered to engulfed structures), rather than radiation delivered to the atmosphere far from the fire.

If the emitting regions occupy all or most of the domain, consider DT and DO.

These comments concern volumetric sources inside the domain, but similar rationale applies to boundary sources, especially when the opacity is small. Thus, in a transparent case, if only a small part of the boundary is hot, ray effects will manifest at other parts of the boundary when DT or DO are used with low resolution.

The combined/hybrid methods are promising, but have not yet been widely adopted in the community and to the author's knowledge have not yet been applied to spectral problems.

Some packages allow collimated radiation in conjunction with diffuse radiation. That enables simulation, for example, of building windows, combining direct sunlight with diffuse radiation. Solar radiation calls for spectral analyses, but depending on how glass or other semi-transparent materials have been modelled, grey simulation can be feasible, as discussed above.

Diffuse and specular behaviour at boundaries is code-dependent, with some packages able to address diffuse behaviour, spectral behaviour, or arbitrary combinations of the two.

Regarding computing costs, typically diffusion models add a negligible amount of CPU, DT and DO add a few percent to tens of percent of CPU (more if finer angular resolution is chosen), and MC adds tens to hundreds of percent of CPU.

5.5   ESTIMATION OF IRRADIATION AT SUB-GRID OBJECTS IN A CFD SIMULATION

A sub-grid object is one which has not been resolved explicitly by the computational mesh. The term is irrelevant to a meshless technology.

A CFD practitioner may decide against resolving certain small objects at the start of a CFD project, or may need to understand the behaviour of small objects which have been added to a design after a complex CFD mesh has been created. Indeed, the practitioner may wish to estimate the radiative loading on relatively small objects at a very large number of locations, or indeed to any point in the computational domain.

In such circumstances, it is legitimate to derive crude estimates of the irradiation at sub-grid entities from the radiation solution delivered by the CFD code, which usually includes the scalar field of mean and incident intensities. By assuming isotropy, the order of magnitude of the local irradiation (W m^{-2}) to a sub-grid body may be derived from

$$q_{Ri} \cong \pi \bar{I} \qquad (5.36)$$

Obviously the errors will be large when the radiation is highly anisotropic (see Section 4.1.8). q_{R_i} can be evaluated at discrete points or may be plotted graphically as contour maps or iso-surfaces.

5.6   COMPUTATIONAL MESHES

A mesh is required by the radiation solver. Since this book deals with the coupling of radiation modelling with CFD, a flow computational mesh will exist, unless the user is employing a meshless CFD code (see Chapter 2). Some CFD packages offer a radiation mesh which is developed from the CFD mesh by grouping flow cells as clusters to form radiation cells. Typically, the user specifies a guide for the number of flow cells to make up the cluster, such as n^3, where $n = 2, 3,$ or 4 say. Sometimes the user may wish to use $n = 1$. Smaller n's improve accuracy but of course increase CPU and cost.

6 Quality Assurance

In order to understand the uncertainty in the results, the computational fluid dynamics practitioner must of course have access to the code's relevant verification and validation. Verification involves comparison of the code's results against exact mathematical or highly accurate numerical benchmark solutions, usually in a 'separate-effects' context, in which the assessment focuses on specific elements of the many physical phenomena which are involved in heat transfer. Validation involves comparison of the code's results against experiments. When validating CFD software and modelling against experiments, it is important for the CFD practitioner to be aware of the experiment scope and measurement methodology. A classic example is temperature measurement, in which unshielded thermocouples can sustain large errors due to radiation (see Section 5.1.1).

Software providers generally subject the software to such tests and share them with their clients. The practitioner must of course check that those assessments are relevant to the scenario that is to be analysed, and that conditions in that scenario are within the range of validity of the tested models.

The CFD practitioner is no less important than the software, and the 'black box' approach, in which the tool and its user are regarded as perfectly reliable, is unwise. The user should possess the ability to understand and formulate the project's needs, match them with the software, modify the modelling if necessary, and assess the results critically. Relevant examples of QA standards and best practice guidelines, at CFD and broader levels, may be found in (Casey and Wintergerste, 2000; Oberkampf *et al.*, 2002; Coleman *et al.*, 2009; Franke *et al.*, 2011; Roe, 2016).

7 Examples

Four examples are provided here briefly. Three involve application to real engineering cases, and one is a laboratory-scale modelling test. Two of the examples deal with grey methods and two with non-grey approaches. The use of spectral methods continues to grow, but in real-life engineering application computing costs and run time are often a limiting factor, and multigrey methodology is a popular compromise nowadays, even if grey models are still being used sometimes.

7.1 UTILITY BOILER

This section briefly describes some analyses of a 300 MWe front-wall-fired pulverised-coal utility boiler. More comprehensive description of the work been reported in (Krishnamoorthy *et al.*, 2010; Nakod *et al.*, 2013), where further details may be found.

The work described here was carried out approximately a decade ago, and pressure to decarbonise notwithstanding, fossil-fired plants of course still exist in large numbers as of 2022. Efforts to reduce the negative impact of coal combustion in existing equipment, other than shutting the plant down, revolve around Clean Coal Technologies (CCT), which have been evolving and been used for many years. Computational fluid dynamics (CFD), as well as physical measurements, are vital for understanding the complex processes which take place in coal combustion, and for mitigating against its environmental consequences.

Oxy-fuel combustion is one of the CCTs which has been considered, and involves removing the nitrogen from the supply air, so that the gas supply is almost pure oxygen, which leads to higher peak flame temperatures. This reduces the quantity of flue gas (consisting mostly of CO_2 and H_2O), as well as the nitrogen oxides (NOx) emissions. Special measures are needed to protect the structure from the higher temperatures, typically by recycling CO_2 into the oxygen supply, and separating the nitrogen requires significant power. It appears that the composition of the ash by-product is generally unaffected by oxy-fuel.

The simulations aimed to do the following:

- Analyse the normal design with air supply.
- Compare results with several Weighted Sum of Grey Gases (WSGGs).
- Analyse oxy-fuel operation.
- Determine NOx emissions.

DOI: 10.1201/9781003168560-7

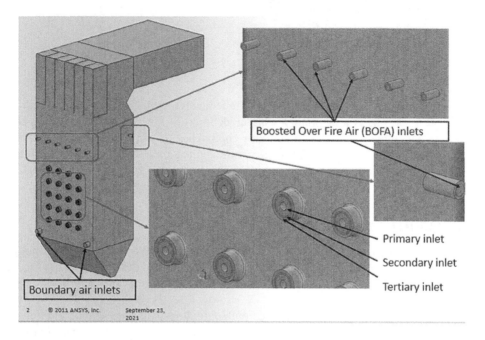

FIGURE 7.1 The coal-fired boiler. The pulverised coal enters through the primary inlet of each nozzle, with secondary and tertiary oxygen supply entering through annular supplies, often with swirl. © Copyright ANSYS.

A schematic of the geometry is shown in Figure 7.1. A selection of some of the operating conditions follows:

- Primary air: 305 ton/h, at 75°C.
- Secondary + tertiary air: 575 ton/h, at 377°C.
- Wall: 452 K, emissivity 0.6.
- Coal: 110 ton/h, high heating value 28.9 MJ/kg.
- Oxidiser compositions: Air 23.3% O_2, 76.7% N_2. Oxy 23.3% O_2. 76.7% CO_2.

The modelling was performed with ANSYS Fluent and used the standard k-ε turbulence model, the Eddy Dissipation combustion model with two reaction steps, and radiation. Six radiation property models were tested (the name Perry does not refer to the originators of the work but rather to the handbook in which they appear, based on the Hottel charts and the SNB RADCAL database, see Hottel *et al.*, 2007 and the two references cited at the start of this section): Smith grey model (Smith *et al.*, 1982), Perry grey model (Hottel *et al.*, 2007), Smith non-grey (4-band, Smith *et al.*, 1982), Perry non-grey (4-band), Perry non-grey (5-band), Chalmers University non-grey (5-band, based on EM2C SNB, Soufiani and Taine, 1997). Some contour maps are shown in Figures 7.2–7.4, and comparisons with temperature and NOx measurements in the real plant rather than in the laboratory (Costa and Azevedo, 2007) are shown in Figures 7.5 and 7.6.

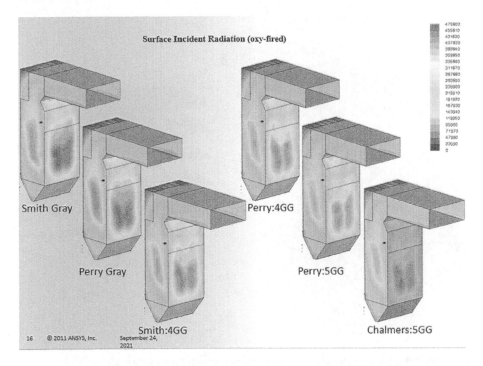

FIGURE 7.2 Surface incident radiative flux, W m⁻², oxy-fuelled. © Copyright ANSYS.

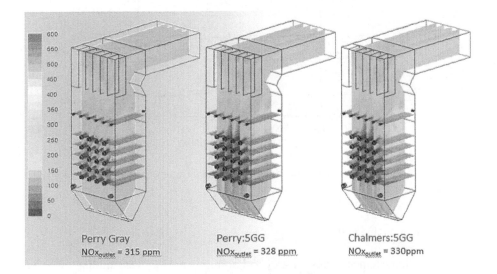

FIGURE 7.3 NOx predictions. © Copyright ANSYS.

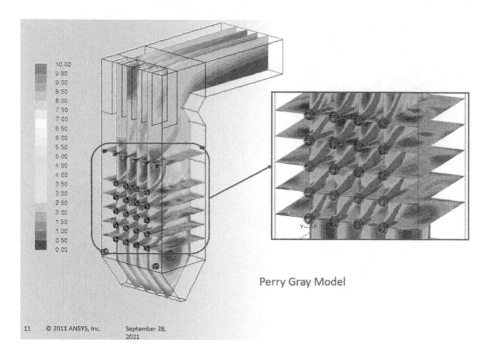

FIGURE 7.4 Ratio of particle to gas absorption coefficients. © Copyright ANSYS.

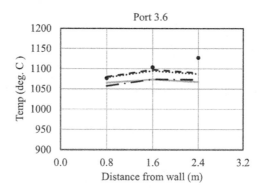

FIGURE 7.5 Comparison with measurement of temperature, air-fuelled. © Copyright ANSYS.

The work concluded that

- Peak flame temperatures for grey and non-grey models were similar.
- Non-grey models predicted radiative fluxes which were about 10% lower than the grey model.
- The number of bands did not significantly affect radiation incident on walls.
- The assessed radiative property models did not affect the species distribution.

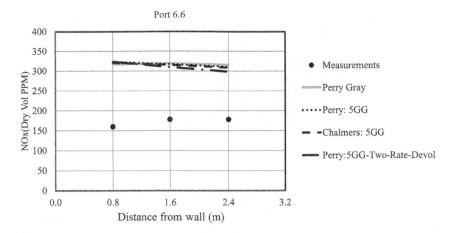

FIGURE 7.6   Comparison with measurements of NOx, air-fuelled. © Copyright ANSYS.

7.2   FORENSIC INVESTIGATION OF A FURNITURE STORE FIRE

It is not absolutely clear when CFD was first used during a forensic fire investigation, but an early example was the fire investigation during the formal government enquiry (Fennel, 1988) into the King's Cross Fire, which had occurred on 18th November 1987. The modelling (Simcox *et al.*, 1988, 1992) employed software which was an early version of that which would later be known as CFX. As an aside, the simulations identified the 'Coanda' phenomenon as the cause for the rapid spread of the flame along the inclined escalator tunnel towards the ticket hall at the top. That eponymous phenomenon is the tendency of jets or plumes to adhere to nearby surfaces, due to low pressures induced by the jet's entrainment of fluid around it, and can therefore influence radiant fluxes delivered to structures too, in addition to convective fluxes. The phenomenon, named the 'trench effect' in the King's Cross enquiry, can also occur without a trench and arises in many fire situations, such as inclined surfaces generally above a critical angle of inclination (e.g. mountain forest fires), and the major influence of some pool shapes on the geometry of pool fire plumes when subjected to cross-wind (Owens and Sinai, 1994; Sinai and Owens, 1995).

This section briefly describes an investigation carried out more recently by NIST (Bryner *et al.*, 2011), of a sofa store fire in 2007 which led to firefighter fatalities. Most of the material here is based on the cited report. That work used Fire Dynamics Simulator (FDS), which is CFD software developed by NIST (McGrattan *et al.*, 2021).

FDS, and some other codes, have also been coupled to evacuation modelling software, which simulate the movement of crowds allowing for varying levels of fitness etc. In some cases this coupling is achieved dynamically during the simulation, accounting for cumulative exposure of individuals to heat and toxins, but it is easier to simulate the crowd movement separately, post-CFD, if that is realistic. Fire safety engineering, of which CFD is just one component, is a complex and eclectic topic.

Sources of techniques and guidelines on fire safety engineering are available in the literature, e.g. (Hurley *et al.*, 2016).

The team investigating the sofa store fire concluded that the fire began outside an enclosed loading area, then reached the loading area, and spread into the retail showroom as well as warehouse spaces. The spread in the two latter areas was hindered by a limited supply of oxygen, leading to a fuel-rich environment, especially in the interstitial space above a suspended ceiling in the main showroom. Eventually fresh air entered these areas through windows at the front of the store, leading to further spread and intensification, including the east and west showrooms. The building layout is shown in Figure 7.7.

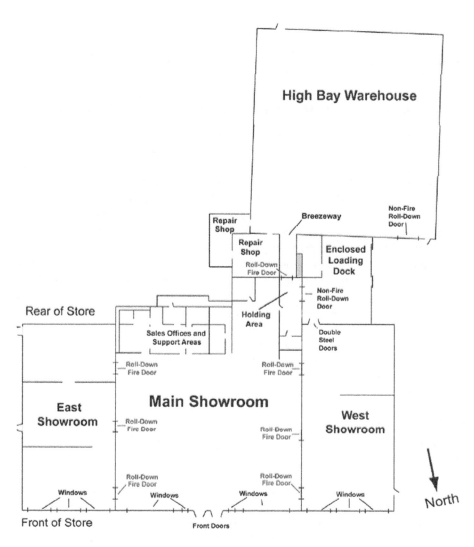

FIGURE 7.7 Floor plan of the sofa superstore. NIST.

The investigating team undertook analyses of five scenarios. The one which will be outlined here is the first one, described as the 'Baseline Scenario', which aimed to simulate the actual event. The others simulated alternative situations which would have occurred if the building design had been different (for example if sprinklers had been installed) or other circumstances had arisen, for example different intervention strategies. That touches on one of the dilemmas which exist in firefighting strategy, with the benefits of smoke removal on the one hand, and the potential risks of introducing more oxygen to a fuel-rich environment on the other. These additional scenarios illustrate the power of CFD, allowing useful lessons to be learned from a forensic project, or indeed during the design phase.

The team compiled a comprehensive catalogue of combustible items (including furnishing, walls, and ceilings) and their burning properties, including energy content. A feel for the scale of the fire is the estimated fuel load of 1450 GJ. The Baseline results were derived from an iterative process between the input data one the one hand, and photos, videos, witness statements, and other documentation on the fire, on the other hand. The runs simulated 40 minutes. Very briefly, the CFD model which was set up included the following:

- The mesh comprised approximately a million cells. This is low by today's standards but was a compromise with the large number of runs which were performed (about 250).
- Turbulence using LES.
- Thermal radiation, absorbing-emitting grey gas. The default model is described below in Section 7.3.
- Assumption that all furnishings were composites of foam and fabric.
- Single-step mixture-fraction exothermic reaction model based on polyurethane foam.
- Tracked products were CO, CO_2 and H_2O.
- Lagrangian particle tracking of water droplets for sprinklers, with heat, mass and momentum exchange between the water drops and the gases.

Flame spread (as opposed to smoke spread) was simulated by computing convection and radiation to surfaces of the fuel elements, modelling transient 1-D conduction within the fuel, triggering ignition when the surface temperature had reached a preset value based on experimental data, and imposing a specified burning rate beyond that time, assuming that all burnt fuel is gasified. In this way, the heat release is computed as a function of time and location, rather than being specified a priori. Figure 7.8 shows the predicted total heat release rate, and Figure 7.9 shows the predicted temperature in a horizontal plane 1.5 m above the floor, at 1560 s.

The simulation of a postulated design including sprinklers indicated that the fire would not spread into the store, with maximum temperatures at sprinkler heads of the order of 70°C (Figure 7.10).

This completes the brief outline, and much more information is of course available in the cited report, which is available online at NIST's website.

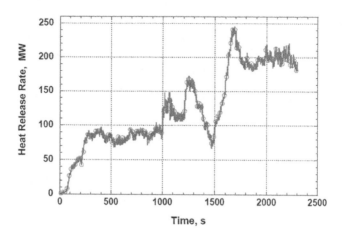

FIGURE 7.8   Baseline total heat release rate. NIST.

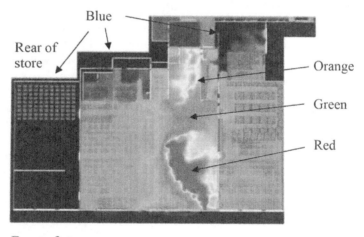

FIGURE 7.9   Temperature map 1.5 m above the floor, Baseline case, $t = 1560$ s. The colours range from 20°C (blue) to 820°C (red). NIST.

7.3   SENSITIVITY TESTS OF GREY GAS MODELS FOR POOL FIRES

This section was contributed by G.C. Fraga. It summarises parts of the material in (Silva *et al.*, 2020; Fernandes *et al.*, 2021), and more details can be found in those papers.

A common modelling approach for radiative transfer in fire simulations is to assume the participating medium to be grey, abbreviated as GG. Its main advantage is that it foregoes the need of solving the RTE multiple times for each optical path, as is required in spectral models. Although known to have poor accuracy, particularly in weakly sooting media, e.g. (Liu *et al.*, 1998; Goutiere *et al.*, 2000), the GG model has

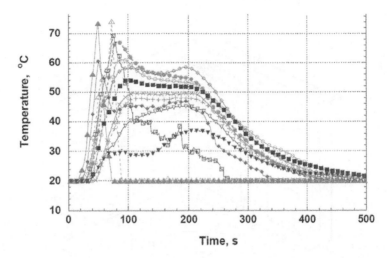

FIGURE 7.10 Furniture store fire. Simulation for a postulated store design including sprinklers. The curves are the temperature at the locations of various sprinkle heads in the store. NIST.

been widely used by the combustion and fire modelling community due to its low computational cost, which is advantageous for coupled calculations, where radiation is to be solved alongside other complex and difficult-to-model processes such as combustion, turbulence and multiphase flow. Coupled fire simulations employing the GG models can be found in (Hostikka *et al.*, 2003; Snegirev, 2004; Krishnamoorthy *et al.*, 2005), for example, and earlier examples exist in the industrial combustion and other fields. The first of the above carried out numerical calculations using the FDS solver for methanol pool fires, with radiation modelled via FDS's default grey gas model, and results for the radiative heat flux and flame temperature showed a good agreement with experimental data. The same pool fires were later studied in the second of those references, now using a different GG model, but only moderate concordance with measurements was observed. More extensive comparisons of simulations to experiments were made in the third paper, for fires both in stagnant air and subjected to cross-wind; the high soot concentration of the configurations was probably a major contributor to the remarkable accuracy which was achieved.

Here, we present results of coupled fire simulations for three scenarios: a closed compartment, an open environment, and a vented compartment. Three common formulations of the GG model (i.e., different approaches for determining the grey absorption coefficient) are tested and compared to experimental data for the radiative heat flux. Focus is given to flames with weakly-to-intermediate soot yields, since, as noted above, there is a reasonable consensus on the applicability of the GG model for strongly sooting flames.

Three fire scenarios are considered, chosen as to cover different types of fires, fuels and heat release rates. These scenarios are based on previous numerical and experimental investigations (Lin *et al.*, 2010; Tu *et al.*, 2013; Sahu *et al.*, 2015) (Figure 7.11).

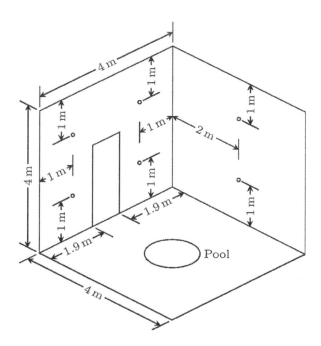

FIGURE 7.11 The geometry of Case 3.

7.3.1 Physical and Numerical Modelling

The cases are simulated in the FDS solver (McGrattan *et al.*, 2012). FDS solves the transport equations for species, momentum and energy for a three-dimensional, compressible, low-Mach number flow; an additional Poisson equation for the perturbation pressure is also solved for the pressure-velocity coupling. Further details on the information quoted here may be found in (McGrattan *et al.*, 2021). Turbulence is modelled with LES, using the Deardorff model (Deardorff, 1980) to close the filtered momentum equation, whereas the filtered species and energy transport equations are closed invoking a standard gradient assumption with constant Prandtl and Schmidt numbers (Poinsot and Veynante, 2005). Detailed discussions can be found in many books on the subject, e.g. (Lesieur *et al.*, 2005; Sagaut, 2006; Garnier *et al.*, 2009). Combustion is modelled as a single-step, mixture-controlled reaction of fuel and air, with the filtered reaction rate approximated via the eddy dissipation concept model (Magnussen and Hjertager, 1977; Poinsot and Veynante, 2005). A lumped species approach, in which the aforementioned transport equations are only solved for the fuel, the oxidant, and the mixture of combustion products, is used in order to reduce the computational cost (Fox, 2003). Fuel evaporation is determined by specifying a constant mass loss rate per unit of burner area, and soot formation is modelled by setting a fixed soot yield, given as the fraction of the burnt fuel mass that is converted into soot volumetrically (McGrattan *et al.*, 2021). The values of mass loss rate and soot yield chosen for each simulation are provided in Table 7.1, and were based on the experimental data cited above.

TABLE 7.1
Summary of Cases

	Closed Compartment	Open Environment	Compartment with an Opening
Base case	Tu *et al.* (2013)	Lin *et al.* (2010)	Sahu *et al.* (2015)
Fuel	Ethanol	Heptane	Methanol
Pool	Squared (0.3 m, 14 mm deep)	Circular (0.3 m diameter, 2 mm deep)	Circular (0.6 m diameter, 120 mm deep)
Mass loss rate (g/m² s)	18.0	20.5	15.5
Soot yield (g soot per g fuel)	0.0012	0.0085	0.0060
Radiant fraction	0.200	0.305	0.200

The radiative transfer equation is solved numerically using the finite volume method (FVM) (Raithby and Chui, 1990). The RTE to be solved has the following form:

$$\frac{dI}{ds} = K_a \left(CI_b - I \right) \tag{7.1}$$

where K_a is the absorption coefficient of the grey medium and C is a source correction factor that is discussed later. To determine K_a, three formulations are considered:

- *FDS's default model:* the absorption coefficient is given as the minimum of two quantities. The first quantity is the Planck-mean absorption coefficient, Equation (5.24), using RADCAL (Grosshandler, 1993) in terms of H_2O, CO_2, and soot. The second quantity is an effective absorption coefficient K_{aE}, obtained by solving the following implicit equation:

$$I_{slb} = \left(1 - e^{-K_{aE}S} \right) I_B + e^{-K_{aE}S} I_{B,slb} \tag{7.2}$$

- This equation describes the intensity I_{slb} leaving a uniform gas layer of thickness S and bounded in one side by a black wall at a temperature $T_{slb} = 900$ K. It is assumed that the gas layer is at the same temperature as the local thermodynamic state of the medium in the actual CFD solution, so it is at that temperature that I_B is evaluated; $I_{B,\,slb}$, on the other hand, is evaluated at T_{slb}. The gas layer width S is taken as five times the size of the local grid cell, and I_{slb} is calculated from a separate non-grey solution using the RADCAL narrow-band model. More information on this grey gas formulation can be found in FDS's technical reference guide (McGrattan *et al.*, 2021).

- K_{aP} - based model: The absorption coefficient is set as equal to the Planck-mean absorption coefficient, estimated according to the correlations of (Cassol *et al.*, 2015). These correlations give K_{aP} as a simple polynomial in temperature, with a linear relation to the concentration of each species, and were determined by fitting Equation (5.24) to the high-resolution spectroscopic database HITEMP2010 (Gordon *et al.*, 2017).
- Cell-based WSGG model: the absorption coefficient is computed from Equation (5.19), with the beam length or thickness again taken as five times the grid cell size. The total emittance ε in the above equation is defined as

$$\varepsilon_g = \frac{1}{I_B} \int_0^\infty I_{B\lambda} \left(1 - e^{-K_{a\lambda}S}\right) d\lambda \tag{7.3}$$

(from which Equation (5.19) can be reached if the medium is assumed to be grey), and is approximated here using the superposition WSGG model (Cassol *et al.*, 2014). As discussed in Section 5.3.1, multigrey methods avoid this problem with a relatively modest increase in cost compared with the cost of spectral methods.

The C parameter in Equation (7.1) is an emission source term correction introduced to account for the fact that FDS's spatial grids often have limited resolution. Consequently, the temperatures computed by the solver tend to be bulk averages, and so can be significantly smaller than the maximum temperature at the flame region. The factor C, which is larger than 1, then artificially increases the local radiative emission to counteract this effect. This factor depends on the estimated radiant fraction of the flame (whose value for each case tested here is also reported on Table 7.1), but its calculation is rather extensive and is not reported here for the sake of conciseness; see (McGrattan *et al.*, 2021). Besides the C parameter, the effect of the small, non-resolved turbulent scales on the radiation field (what is commonly referred to as sub-filter-scale or sub-grid-scale (SGS) turbulence-radiation interaction; see Section 4.4.1) is neglected, and the filtered radiative intensity is taken as equal to that calculated from the filtered temperature and medium composition via Equation (7.1).

The numerical simulations are transient and use a second-order, explicit, predictor-corrector scheme to advance the solution in time, with the time-step size defined dynamically in order to maintain a local Courant-Friedrichs-Lewy number between 0.8 and 1.0. The solution is second-order accurate in space as well, employing a total-variation-diminishing scheme for modelling the convective transport (Roe, 1986). Discretisation of the transport equations is achieved with rectilinear, non-uniform meshes, with refinement at the flame and radiometers. Gird quality was ensured by assessing the D^* and M criteria (Pope, 2000; Hurley *et al.*, 2016). A total of 500 control angles are employed for the directional integration of the radiative intensity following the FVM; further increasing this parameter showed no appreciable changes in the results.

7.3.2 Results and Discussion

7.3.2.1 Closed Compartment

The first test case reproduces one of the configurations that were experimentally studied by (Tu *et al.*, 2013), which consists of a square-shaped, 0.3 m-wide pool of liquid ethanol burning in air inside a close compartment measuring 10 m × 7 m × 4 m. For the numerical simulations, the computational domain was set to these same dimensions, with the boundaries assumed to be solid, black walls at 293 K. In the experiment a nominally steady state existed between approximately 200 s and 550 s, the simulations ran for a total of 530 s and all data reported next were averaged over the last 230 s of the calculation.

Of major interest here are comparisons of the radiative heat flux, which had been measured by two radiometers positioned 1.5 m from the centre of the pool in orthogonal directions. Comparing the simulation results with the experimental data, both the default FDS grey gas model and the model based on the WSGG-calculated emittance are capable of predicting the radiative heat flux within 10% of the measurements; furthermore, while both overestimate the quantity, the model based on the Planck-mean absorption coefficient underestimate the heat flux by as much as 15%. Simulations were also carried out both without soot formation and excluding the source correction factor C in the RTE, and only small differences were observed in the radiative fluxes (Silva *et al.*, 2020; Fernandes *et al.*, 2021). The measured fluxes at the radiometers were both approximately 473 W m^{-2}.

7.3.2.2 Open Environment

This second case is based on the circular heptane pool fire studied by (Lin *et al.*, 2010), burning with air in an open environment at atmospheric conditions. For the simulations, the computational domain was defined as a 1.5 m × 1.2 m × 2.5 m box, with the fuel pool positioned at the centre of the bottom surface. Once again, all results presented here correspond to averages of the instantaneous data computed over an interval where the solution is in a statistically steady state (in this case, between 100 s and 300 s after the simulation start).

The incident radiative heat flux 0.75 m away from the centreline of the flame had been measured by radiometers positioned at different heights above the pool surface. In contrast with the closed-compartment case, presently all GG models underpredict the heat flux, although again the K_{aP}-based model has the worst performance; for the WSGG-based GG model, the results fall within the uncertainty bounds of the measurements (see Figure 7.12). It is worth noting, however, that only marginal differences were observed for the predicted temperature and concentration fields between the three GG models tested in (Fernandes *et al.*, 2021), despite the fairly large importance of radiation for that case (as measured by the radiant fraction of the case, cf. Table 7.1).

7.3.2.3 Vented Compartment

Lastly, we consider the scenario studied experimentally by (Sahu *et al.*, 2015), involving a methanol pool fire within a compartment with an opening. The enclosure is a cubic structure with side 4 m, and the opening measures 2.0 m × 0.2 m. In the

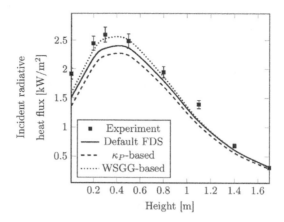

FIGURE 7.12 Case 2, comparison of sooted predictions with experiment. Adapted from (Fernandes *et al.*, 2021).

present simulations, extended the domain 0.5 m beyond the walls of the compartment on each side and an opening condition was imposed at all those boundaries; see the discussion in Section 5.1 and Figure 5.1. Heat transfer through the structure was modelled as one-dimensional heat conduction, with walls assumed to be made of bricks for the calculation of the thermal conductivity.

According to (Sahu *et al.*, 2015), the relatively small size of the containment structure (compared to the fuel pool, that measures 60 cm in diameter, cf. Table 7.1), delays the establishment of a statistically steady state, and in fact no such state was observed for the entire duration of the experiment, which lasted 1800s. Therefore, only transient data is reported there, and the comparisons with the numerical results presented next are based on these. Notice also that the rig measured the net heat flux, rather than the radiative heat flux (as in the previous comparisons), which was done through sensors positioned at the compartment walls, 1 m and 3 m high. The work compared the running averages of the instantaneous data over a 20 s time interval (Figure 7.13).

The simulations do not show the same good level of agreement as they did for the previous cases; the error of the numerical results, averaged over the entire 1000 s duration of the simulations, is about 10%, but instantaneously this error can be as high as 35%. It should be noted, however, that these large errors may not be due to the GG model necessarily, since the comparisons include inaccuracies associated to the numerical determination of the convective heat flux as well (which in FDS is done through a correlation for combined forced and natural convection); furthermore, errors of 35% are still within bounds of what is expected for GG simulations of weakly sooting flames—see, for example (Liu *et al.*, 1998; Goutiere *et al.*, 2000). Finally, notice that, differently from the other cases, here the K_{aP}-based formulation does not consistently provide the worst predictions, and even gives the closest predictions at the highest measuring point.

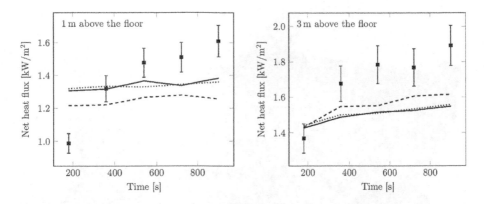

FIGURE 7.13 Case 3, vented compartment – comparison of net wall heat fluxes (Fernandes *et al.*, 2021).

7.3.3 FINAL REMARKS

Here we presented some illustrative examples of the solution of real fire scenarios where the contribution of radiative transfer was accounted for by solving the RTE. Coupled simulations were compared with experimental data for open, confined and partially confined fires above pools of different fuels. Only models that assume the participating medium to be grey were tested, with one RTE being solved, since they are computationally more efficient than non-grey models and retain some accuracy especially for high soot loads. Even in the cases tested here, where the soot production was weak to moderate, the models still performed fairly well, particularly for the open-compartment and closed-compartment cases, where errors for the predicted radiative heat flux were below 10%.

7.4 HEADLIGHT

This section was provided by Mentor, a Siemens business.

In this study a conjugate heat transfer analysis of an automotive headlamp is considered, including convection, conduction and radiation. One of the goals of the study is to model surface condensation on the inside face of the front lens using a thin-film model for the film condensation. Since the formation of the condensation film is highly dependent on the temperature field, accurate modelling of the thermal effects is necessary which includes conjugate heat transfer inside the headlight within a single simulation. The CFD model uses a low Reynolds Number variant of the k-epsilon model as the bulk turbulence model for the internal air with a bespoke wall function treatment. Thermal conduction was included within all the solid materials in the headlight assembly, and includes radiation as an important contributor. The external ambient conditions are imposed onto the outside surfaces of the headlight through heat transfer coefficients to the external ambient temperature of −5°C.

A transient analysis was conducted simulating a "soak" condition of a vehicle having been parked and cooling in a cold ambient environment, followed by a period

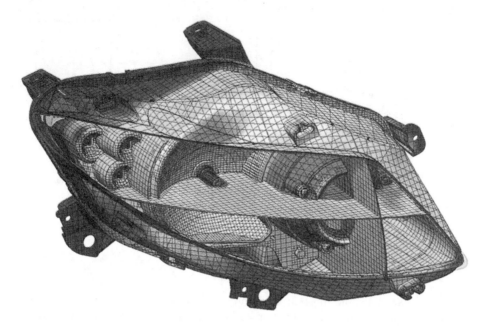

FIGURE 7.14   Surface plot of temperature contours and mesh at a selected instant. © Siemens Digital Industries Software.

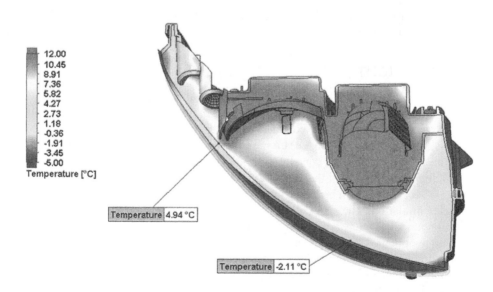

FIGURE 7.15   Temperature contours on a horizontal slice through the headlamp assembly. © Siemens Digital Industries Software.

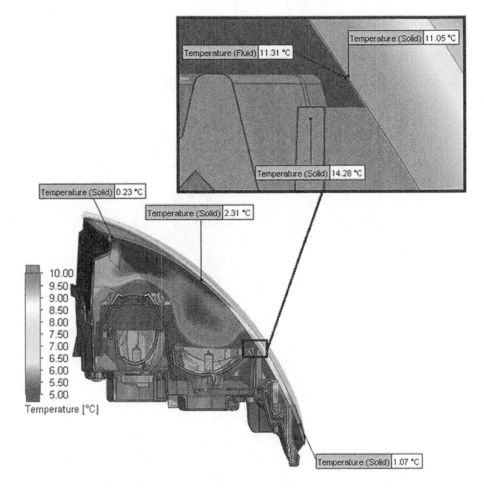

FIGURE 7.16   Detail of temperature difference between headlamp lens and adjacent fluid.
© Siemens Digital Industries Software.

in which the lights are turned on. When the lights are engaged the front lens is heated
in a non-uniform distribution based on the radiative properties of the relevant parts
(in particular the bulbs, reflectors and lenses) and the material properties and geometry of the headlamp assembly. The geometry of the headlamp is used directly from
the computer-aided design (CAD) system and not simplified prior to meshing or
simulation (Figures 7.14–7.16).

The Monte Carlo method was used for radiation modelling in this analysis. The
main idea of this approach is to emit a large number of photons from randomly
selected locations on a given surface or in a given volume and then to track their
progress through a series of interactions with media and boundaries until its 'weight'
falls below some minimum at which point it 'dies'. The properties of the enclosure

and the laws of probability are used to simulate an interaction with a given surface. This method was chosen because of its accurate handling of the relevant radiative considerations for this problem, including:

- Absorption, transmission in semi-transparent solids
- Reflection
- Refraction
- Spectral characteristics: Properties like absorptivity and emissivity with complex wavelength dependencies are modelled with a convenient "ray-based" approach suited to the Monte Carlo method.

Appendix A

DIMENSIONLESS GROUPS

There are many dimensionless groups within the disciplines of fluid dynamics and heat transfer, and a selection is given here (Table A.1), albeit an incomplete one, and fuller sets are available in the literature. The symbol Δ denotes the modulus of the difference between two values, representing a driving force (which might not be the only one) behind the scenario, for example temperature difference between a wall and the ambient far from it, or the vapour concentration difference between a liquid free surface (i.e. a liquid–gas interface) and the gaseous ambient well away from the interface.

TABLE A.1
Selection of Dimensionless Groups

Symbol	Dimensionless Number Name	Expression	Comments
Bi	Biot	$\dfrac{hL_S}{k_S}; L_s = \dfrac{V_S}{A_S}$	External thermal resistance/internal thermal resistance of solid
Bo	Boltzmann	$\dfrac{\rho C_P U}{\varepsilon \sigma T_w^3}$	Advection/radiation. Also known as the Thring number N_{Th}
Bo	Bond	$\dfrac{g\left(\rho_L - \rho_G\right)L^2}{\sigma}$	Gravitational/surface tension forces
Br	Brinkman	$\dfrac{\mu U^2}{k\Delta T}$	Viscous dissipation/thermal conduction
C_f	Friction coefficient	$\dfrac{2\hat{\tau}_w}{\rho U^2}$	Frictional/dynamic forces
Ec	Eckert	$\dfrac{U^2}{C_p \Delta T}$	Kinetic/wall convective energy
Fr	Froude	$\dfrac{U}{\sqrt{gL}}$	Square root of dynamic/gravitational
$\hat{Fr}$	Densimetric Froude	$\left(\dfrac{\rho_o U^2}{g\Delta\rho L}\right)^{1/2} = \dfrac{1}{\sqrt{Ri}}$	Generalisation of Froude number

(Continued)

TABLE A.1 (Continued)

Symbol	Dimensionless Number Name	Expression	Comments
j_H	Colburn j factor	$St Pr^{2/3}$	Heat-momentum analogy
Fo	Fourier	$\dfrac{\lambda t}{L_s^2}$	Time/thermal diffusion time in solid
Gr	Grashof	$\dfrac{g\Delta\rho L^3}{\rho_o \nu^2}$	Buoyancy/viscous forces
Ja	Jacob	$\dfrac{C_p\left(T_w - T_{sat}\right)}{h_{LG}}$	Sensible/latent heat. Inverse of the Kutateladze number Ku
Le	Lewis	$\dfrac{\mathcal{D}}{\lambda}$	Mass/thermal diffusivities
M	Mach	$\dfrac{U}{a}$	Flow speed/sound speed
N_R	Radiation	$\dfrac{h\left(T_g - T_w\right)}{\varepsilon\sigma\left(T_g^4 - T_w^4\right)}$	Wall convection/radiation. Also known as the Hottel number
Nu	Nusselt	$\dfrac{hL}{k_g}$	Dimensionless heat transfer coefficient
Pr	Prandtl	$\dfrac{\nu}{\lambda}$	Molecular momentum diffusivity/ thermal diffusivity
Pe	Peclet	$\dfrac{UL}{\lambda} = PrRe$	Dimensionless heat transfer parameter
Ra	Rayleigh	$PrGr$	Modified Grashof number
Re	Reynolds	$\dfrac{UL}{\nu}$	Inertial/viscous forces
Ri	Richardson	$\dfrac{g\Delta\rho L}{\varrho_o U^2}$	Buoyancy/dynamic heads. Also known as Archimedes Number (Ar)
Ro	Rossby	$\dfrac{U}{\hat{\Omega}L}$	Inertial/Coriolis forces
Sc	Schmidt	$\dfrac{\nu}{\mathcal{D}}$	Momentum/mass diffusivities
Sh	Sherwood	$\dfrac{h_m L}{\mathcal{D}}$	Dimensionless mass transfer coefficient
St	Stanton	$\dfrac{h}{\varrho C_p U} = \dfrac{Nu}{Pe}$	Heat transfer coefficient relative to advection

Symbol	Dimensionless Number Name	Expression	Comments
We	Weber	$\dfrac{\rho U^2 L}{\hat{\sigma}}$	Inertia/surface tension
y_+ (or y_*)	Yplus or y-plus	$\dfrac{y}{\nu}\left(\dfrac{\lvert \tau_w \rvert}{\rho}\right)^{1/2}$	Normal distance/viscous scale
τ_R	Opacity	KL	Dimensionless radiation extinction coefficient

Appendix B

THE ELECTRICAL ANALOGY

The basics of this subject, which is the classical engineering method for transparent media, have already been introduced in Section 4.1.5, and its application discussed in Section 5.4.1. This appendix provides more detail. The technique is also known in the literature as the network method. A partially similar approach is the net radiation method, already cited, and of course the radiation solvers available in most computational fluid dynamics (CFD) codes offer the various numerical algorithms.

When constructing the electrical analogue, use is made of what are described in the literature (Holman, 2018) as the 'surface' and 'space' resistances. The former is placed at each surface and the latter between every pair of surfaces which are visible to each other. By a process known as 'view factor algebra', the following set of linear simultaneous algebraic equations are derived, for the radiosity at each of the N surfaces bounding the domain.

$$J_i - \left(1 - \varepsilon_i\right) \sum_{j=1}^{N} F_{ij} J_j = \varepsilon_i E_{bi} \tag{B.1}$$

A total of N^2 view factors are required. However, due to reciprocity and the summation rule, only $N(N-1)/2$ view factors are needed. Extensive catalogues are available, in both graphical and analytical forms, e.g. (Howell *et al.*, 2021).

If the view factors are known, this is a system of N linear simultaneous equations for the N radiosities, which in turn yield the net radiative flux at each surface using Equation (4.17). If the view factors are unknown, then they need to be computed or obtained from the literature. One can in fact use a radiation solver to compute view factors, for example by setting the temperature of a given surface (denoted by j) to $\sigma^{-1/4} = 64.8$ K and all the others to zero Kelvin. The computed irradiation (in Watts) at any surface i is then $A_j F_{ji}$ or $A_i F_{ij}$. Utilisation of the view factors, if the transparent assumption is a valid one, can save much computing effort, although as already pointed out, some radiation models are cheap to run (e.g. Discrete Transfer).

Equation (B.1) assumes that all the surfaces bounding the domain are diffuse. It is possible to generalise this to a combination of diffuse and specular surfaces, e.g. (Holman, 2018; Modest, 2021; Howell *et al.*, 2021).

For the record, it is worth quoting the classical formula for a grey two-surface enclosure, i.e. a transparent enclosure with its envelope made up of only two opaque, grey and diffuse surfaces, each possessing its own area A_i, temperature T_i and emissivity ϵ_i. The net energy flow (W) from surface 1 to surface 2 is given by

$$Q_{12} = \frac{E_1 - E_2}{\dfrac{(1-\epsilon_1)}{\epsilon_1 A_1} + \dfrac{1}{A_1 F_{12}} + \dfrac{(1-\epsilon_2)}{\epsilon_2 A_2}} \tag{B.2}$$

where $E_i = \sigma T_i^4$. This degenerates to a number of classical cases, as discussed in the literature, such as two infinite parallel plates ($A_1 = A_2$) and two infinite coaxial cylinders. The result for the former is the familiar

$$q_{12} = \frac{E_1 - E_2}{\dfrac{1}{\varepsilon_1} + \dfrac{1}{\varepsilon_2} - 1} \tag{B.3}$$

where q_{12} is the heat flux (W m^{-2}). These concepts carry over to participating media, as exemplified by the Zonal method, which involves exchange factor between surfaces, between gas zones, and between gas zones and surfaces. However, such approaches are now outdated.

Appendix C

FRESNEL'S EQUATIONS

The behaviour of atmospheric radiation and its interaction with semi-transparent materials has been an important topic for a long time, and climate change has enforced that situation due to the need to improve buildings' energy efficiency and indoor air quality. Since CFD practitioners in the built environment are likely to model windows, it is worth including the classical Fresnel relationships for reflection and transmission at an optically smooth interface between two isotropic and homogeneous dielectric media, i.e. media which are non-conducting (in the electrical sense) and not opaque on the spatial scales being considered, and in which the absorptive index (see Equation (4.10)) is negligible, e.g. (Modest, 2021; Howell *et al.*, 2021). The results are generalised in the cited literature to account for absorption.

Detailed discussion of polarisation is beyond the scope of this book, and it has already been pointed out that it can generally be ignored for thermal radiation. There are exceptions, however, and a topical example is solar radiation and its interaction with glass and other semi-transparent materials, an issue that is relevant to energy efficiency in the built environment and to solar power. CFD practitioners often need to consider these issues when analysing building performance. Polarisation will be touched on here, but only briefly.

First, a few common terms will be defined: The *wave vector* defines the direction of propagation of a wave. The *plane of incidence* is the plane containing both the incident and the reflected wave vectors (and the normal to the surface). Electromagnetic waves are transverse and are made up of an electric and a magnetic field both of which oscillate perpendicularly to the wave vector. The *polarisation* of a wave is a measure of the characteristics (the orientation) of the electric component of the wave.

The classical Fresnel theory depends on the polarisation state of the wave, and leads to two relationships for the reflection coefficient $\hat{R}$. One is typically named *S-Polarised* or *Perpendicularly Polarised*, and denoted by the subscript '⊥', and the other is named *P-Polarised* or *Parallel Polarised*, and denoted by the subscript '∥'.

In the current notation, a plane wave arrives from medium '1', possessing refractive index n_1 (cf. Section 4.1.3), towards the plane interface separating medium '1' from medium '2' (possessing refractive index n_2). It is assumed initially that $n_2 > n_1$. Define the ratio of refractive indices:

$$\tilde{n} = \frac{n_2}{n_1} \tag{C.1}$$

The classical Fresnel relationships are as follows:
  P-Polarisation (parallel):

$$\hat{R}_{\parallel} = \left(\frac{n_1 \cos\theta_2 - n_2 \cos\theta_1}{n_1 \cos\theta_2 + n_2 \cos\theta_1} \right)^2 \tag{C.2}$$

S-Polarisation (perpendicular):

$$\hat{R}_{\perp} = \left(\frac{n_1 \cos\theta_1 - n_2 \cos\theta_2}{n_1 \cos\theta_1 + n_2 \cos\theta_2} \right)^2 \tag{C.3}$$

When $\theta_1 = 0$ (i.e. at normal incidence), the two reflectivities are equal (already quoted in Section 4.1.9, Equation (4.28). When the incident wave is unpolarised, the arithmetic mean of these two reflectivities may be used, as well as generally for engineering heat transfer.

$$\hat{R}_m = 0.5 \left(\hat{R}_{\parallel} + \hat{R}_{\perp} \right) \tag{C.4}$$

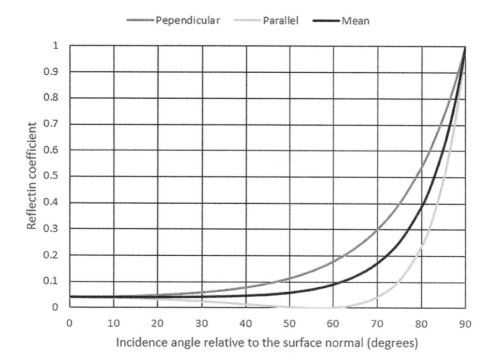

FIGURE C.1 Fresnel reflection coefficients for a plane smooth interface between non-absorbing or weakly-absorbing media, with $n_1 = 1$ and $n_2 = 1.5$. 'Perpendicular' (top curve) refers to the S-Polarisation, 'Parallel' (bottom curve) refers to P-Polarisation, and 'Mean' (middle curve) refers to the arithmetic mean of those two.

These three reflectivities are plotted in Figure C.1 as a function of the incidence angle (i.e. the angle between the incident wave vector and the normal to the interface), for $n_1 = 1$ and $n_2 = 1.5$, which is representative of the air-glass pair. The Brewster angle is 56.3° for this pair, and the critical angle, if the values of the refractive indices are reversed or for waves travelling in the opposite direction (from medium 2 towards medium 1), is 42.0°.

The Brewster angle, at which $\hat{R}_{\parallel}$ vanishes, is given by

$$\theta_{1B} = \sin^{-1}\left(\frac{\tilde{n}^2}{\tilde{n}^2+1}\right)^{0.5} = \tan^{-1}\tilde{n} \tag{C.5}$$

If $n_1 > n_2$, the critical angle beyond which no transmission occurs, is given by

$$\theta_{1C} = \sin^{-1}\tilde{n} \tag{C.6}$$

Appendix D

SPHERICAL COORDINATES, AND MORE ON SCATTERING

First, for spherical coordinates (Figure D.1), the integral of any function $\Upsilon(\theta, \varphi)$ over solid angle is given by

$$\int \Upsilon \, d\Omega = \int\limits_{\phi=0}^{2\pi} \int\limits_{\theta_0}^{\pi/2} \Upsilon(\theta', \phi') \sin\theta' \, d\theta' d\phi' \tag{D.1}$$

where θ_0 equals $-\pi/2$ for integration over a sphere and 0 for a hemisphere.

Turning now to the radiative transfer equation (RTE), Equation (4.41), it is commonplace to assume that the scattering phase function $\tilde{p}$ is only a function of the angle between the incident ray and the direction being considered, i.e. between the unit vectors $\vec{\Omega}'$ and $\vec{\Omega}$.

In spherical coordinates, the unit vector in the direction of $\vec{\Omega}$ (and indeed any vector) is

$$\vec{\Omega} = \begin{bmatrix} \sin\theta\cos\phi \\ \sin\theta\sin\phi \\ \cos\theta \end{bmatrix} \tag{D.2}$$

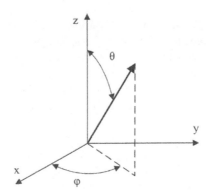

FIGURE D.1 Spherical co-ordinates – azimuthal angle ϕ and polar (or colatitude, zenith) angle θ.

Denote ψ as the angle between $\vec{\Omega}'$ and $\vec{\Omega}$, where the dash relates to the dummy variable of integration in Equation (4.41). Since $\cos(\psi) = \vec{\Omega} \cdot \vec{\Omega}'$, and writing

$$\mu_\psi = \cos(\psi), \mu = \cos(\theta), \mu' = \cos(\theta') \tag{D.3}$$

it follows that

$$\mu_\psi = \mu\mu' + \left(1 - \mu^2\right)^{1/2} \left(1 - \mu'^2\right)^{1/2} \cos(\phi' - \phi) \tag{D.4}$$

The scattering integral is then

$$\int_{\phi=0-\pi/2}^{2\pi} \int^{\pi/2} I_\lambda(\theta',\phi') \, \tilde{p}_\lambda(\mu_\psi) \sin\theta' \, d\theta' \, d\phi' \tag{D.5}$$

The phase function is usually written in terms of Legendre functions, Equation (4.44). Some analyses have included a large number of terms, approaching and exceeding a hundred. The textbooks discuss a number of models for the phase function, and views on their performances vary. The models mentioned commonly are isotropic, linearly anisotropic, Rayleigh, Delta-Eddington, Henyey-Greenstein, and Crosbie-Davidson, e.g. (Modest, 2021; Howell et al., 2021). The classical Rayleigh phase function for particles which are small (on the scale of the wavelength), whatever their shape, can be obtained from $N = 2$ in Equation (4.44) with $a_1 = 0$ and $a_2 = 0.5$:

$$\tilde{p}_\lambda = \frac{3}{4}\left(1 + \mu_\psi^2\right) \tag{D.6}$$

This possesses equal forward and backward scatter peaks. Several attempts have been made to approximate and simplify anisotropic scattering using scaling or similarity laws. The so-called *transport approximation* scales the unit of length leading to an equivalent isotropic model. Other attempts (Lee and Buckius, 1982, 1983; Truelove, 1984) simplify some scattering cases: They reduce an isotropically scattering problem to a non-scattering one, and a linearly scattering one to an isotropically scattering one.

Appendix E

EXACT CLOSED-FORM SOLUTION FOR AN INFINITE, PLANE, GREY, HOMOGENEOUS, ABSORBING-EMITTING SLAB BETWEEN TWO PLATES WITH DIFFERENT TEMPERATURES AND EMISSIVITIES

This section gives an example which provides a useful baseline for software verification, as well as an illustration of the complexity of thermal radiation. It is one of the rare cases where an exact closed-form analytical solution can be derived: An infinite, plane, homogeneous, isothermal (at temperature T_g), grey, absorbing-emitting slab. The slab lies between two diffuse, grey parallel plates which possess different emissivities and are at different uniform temperatures. Note that this case does not feature radiative equilibrium, but is rather a non-equilibrium example, and is relevant to realistic scenarios with combined heat transfer in which, for example, convection plays a role too (e.g. Bénard convection at high Rayleigh numbers). Some other exact closed-form cases can be found in the cited literature, e.g. (Modest, 2021; Howell *et al.*, 2021), and many papers have been published on approximate and numerical solutions of this case.

From the mathematical perspective, because the problem excludes scattering and the temperature field is specified rather than calculated, the radiative transfer equation (RTE) emerges as a differential equation rather than an integral equation, which of course simplifies matters.

Denote the thickness of the slab by L, its absorption coefficient by K_a, and assume that its normal is parallel to the z axis, with one boundary at $z = 0$ and the other at $z = L$. A flux is positive when it is in the positive z direction, i.e. in the direction of surface 1 to surface 2.

Consider first boundaries which are at absolute zero and are perfectly black. The classical distribution of radiative flux across the slab is then derived from the grey versions of Equations (4.37) and (4.41) and integrating along a ray normal to the slab, leading to two coupled integral equations for the local radiative flux and the local incident radiation. In the non-scattering case this yields (cf. Modest, 2021; Howell *et al.*, 2021):

$$q_R(\tau) = 2\big[E_3(\tau_L - \tau) - E_3(\tau)\big]e_g \qquad \text{(E.1)}$$

where E_3 is the Exponential Integral (Abramowitz and Stegun, 1965), with $n = 3$:

$$E_n(z) = \int_1^\infty \frac{\exp(-zt)}{t^n} \, dt \tag{E.2}$$

Note that the Exponential Integral can be expressed in terms of the Gamma Function Γ, which can be convenient when evaluating such functions in a spreadsheet, for example.

$$E_n(z) = z^{n-1}\Gamma(1-n,z) \tag{E.3}$$

Also

$$e_g = \sigma T_g^4; \tau = K_a z; \tau_L = K_a L \tag{E.4}$$

Thus τ varies between 0 and τ_L. Equation (E.1) essentially defines the gas-boundary view factor and the slab emissivity for this case, say by putting $\tau = \tau_L$:

$$\epsilon_g = 1 - 2E_3(\tau_L) \tag{E.5}$$

Figure E.1 depicts this emissivity as a function of the opacity.

Now consider the case where the boundaries are diffuse walls at finite temperatures T_1 and T_2, and possess emissivities ε_1 and ε_2 respectively. Invoking the boundary condition Equation (4.17) at both walls, and exploiting the fact that $E_3(0) = 0.5$, the radiative flux (as a function of z) is now, after some algebra,

$$q_R = 2(J_1 - e_g)E_3(\tau) - 2(J_2 - e_g)E_3(\tau_L - \tau) \tag{E.6}$$

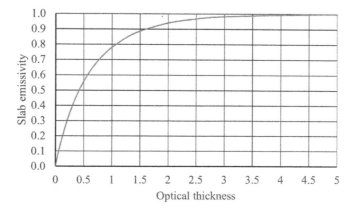

FIGURE E.1 Exact theory for the plane grey, homogeneous, absorbing-emitting plane slab – the slab emissivity ϵ_g vs. the opacity τ_L, black and cold boundaries.

The two radiosities at the boundaries 1 and 2 are

$$J_1 = \hat{\Delta}^{-1}\left\{\varepsilon_1 e_1 + 2\hat{\alpha}\varepsilon_2\left(1-\varepsilon_1\right)e_2 + \hat{\beta}e_g\left[1-\varepsilon_1+2\hat{\alpha}\left(1-\varepsilon_1\right)\left(1-\varepsilon_2\right)\right]\right\} \qquad \text{(E.7)}$$

$$J_2 = \hat{\Delta}^{-1}\left\{2\hat{\alpha}\epsilon_1\left(1-\epsilon_2\right)e_1 + \epsilon_2 e_2 + \hat{\beta}e_g\left[1-\epsilon_2+2\hat{\alpha}\left(1-\epsilon_1\right)\left(1-\epsilon_2\right)\right]\right\} \qquad \text{(E.8)}$$

where

$$\hat{\Delta} = 1 - B_1 B_2 \hat{\alpha}^2; \quad B_1 = 2\left(1-\varepsilon_1\right); \quad B_2 = 2\left(1-\varepsilon_2\right) \qquad \text{(E.9)}$$

$$\hat{\alpha} = E_3\left(\tau_L\right); \quad \hat{\beta} = 1 - 2\hat{\alpha}; \quad e_1 = \sigma T_1^4; \quad e_2 = \sigma T_2^4 \qquad \text{(E.10)}$$

The fluxes at surfaces 1 and 2 are given explicitly by

$$q_{R_1} = J_1 - e_g - 2\left(J_2 - e_g\right)E_3\left(\tau_L\right) \qquad \text{(E.11)}$$

$$q_{R_2} = 2\left(J_1 - e_g\right)E_3\left(\tau_L\right) - J_2 + e_g \qquad \text{(E.12)}$$

The related examples given in (Modest, 2021; Howell *et al.*, 2021) are for black boundaries or boundaries possessing the same temperature and emissivity. The present results appeared twice previously (Williams, 1983, 1984; Ozisik, 1985).

The heat flux delivered to surface 2 may alternatively be written in a format reminiscent of the network method:

$$q_{R2} = \sigma F_{12}\left(T_1^4 - T_2^4\right) + \sigma F_{2g}\left(T_g^4 - T_2^4\right) \qquad \text{(E.13)}$$

where

$$F_{12} = 2\hat{\alpha}\epsilon_1\epsilon_2 / \hat{\Delta} \qquad \text{(E.14)}$$

and

$$F_{2g} = \hat{\beta}\epsilon_2\left[1 + 2\hat{\alpha}\left(1-\epsilon_1\right)\right] / \hat{\Delta} \qquad \text{(E.15)}$$

Similarly, the flux at surface 1 can be written as

$$q_{R1} = \sigma F_{12}\left(T_1^4 - T_2^4\right) + \sigma F_{1g}\left(T_1^4 - T_g^4\right) \qquad \text{(E.16)}$$

where

$$F_{1g} = \hat{\beta}\epsilon_1 \left[1 + 2\hat{\alpha}\left(1 - \epsilon_2\right)\right] / \hat{\Delta} \qquad\qquad (E.17)$$

These expressions acquire the familiar forms in the optically thin (transparent) and optically thick limits, viz. $\tau_L \to 0$ and $\tau_L \to \infty$, respectively. In the latter case, neither boundary can 'see' the other, and the heat flux at a boundary is identical to a transparent case involving that boundary and another at temperature T_g with emissivity 1.0.

Appendix F

EXACT CLOSED-FORM SOLUTION FOR AN INFINITE, PLANE, GREY, ABSORBING-EMITTING SLAB IN RADIATIVE EQUILIBRIUM BETWEEN TWO PLATES WITH DIFFERENT TEMPERATURES AND EMISSIVITIES

As discussed elsewhere in this book, the radiative equilibrium situation tends to exist when thermal radiation is much larger than the other forms of heat transfer. The temperature field is dominated by the radiative field and is part of the solution. The exact solution for the present case was first published by (Heaslet and Warming, 1965), which also accounted for a volumetric heat source within the medium. Their result for the heat flux excluding that source is reproduced here, and the reader is referred to (Heaslet and Warming, 1965; Modest, 2021; Howell *et al.*, 2021) for the temperature distribution across the slab. It should be noted that the theory predicts a temperature 'slip' or discontinuity at the walls. The flux is given by

$$q_R = \frac{\left(e_1 - e_2\right)}{\left(\dfrac{1}{\varepsilon_1} + \dfrac{1}{\varepsilon_2} - 2\right) + \dfrac{1}{\Psi_b}} \tag{F.1}$$

Here Ψ_b is a function of τ_L (cf. Equation (E.4)) and depends on the so-called Chandrasekhar or Ambartsumian functions (Heaslet and Warming, 1965). This will not be detailed here, and instead the values of Ψ_b will be tabulated and plotted (Table F.1 and Figure F.1).

This result should be compared with the transparent case, Equation (B.3). An important point to note is that the heat flux in this equilibrium situation cannot be larger than that which would arise in the transparent case. This contrasts with the non-equilibrium situation, where a potential exists for the flux to be either smaller or larger than the transparent flux due to the influence of other modes of heat transfer, and sources thereof.

TABLE F.1
Equilibrium Chandrasekhar/Ambartsumian
Number vs. Opacity

τ_L	Ψ_b	τ_L	Ψ_b
0	1.0000	2.0	0.3900
0.1	0.9157	3.0	0.2918
0.2	0.8491	4.0	0.2394
0.3	0.7934	5.0	0.2030
0.4	0.7458	6.0	0.1797
0.5	0.7040	7.0	0.1583
0.6	0.6672	8.0	0.1415
0.8	0.6046	9.0	0.1279
1.0	0.5532	10.0	0.1167
1.5	0.4572		

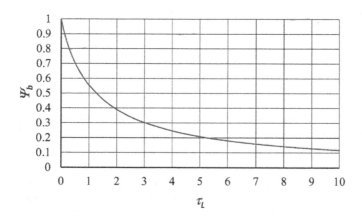

FIGURE F.1 Ψ_b vs. opacity for equilibrium radiation across a grey plane-parallel slab.

Appendix G

THE WILLIAMS THEORY FOR AN INFINITE, PLANE, GREY, HOMOGENEOUS, ABSORBING-EMITTING, ISOTROPICALLY SCATTERING SLAB BETWEEN TWO PLATES WITH DIFFERENT TEMPERATURES AND EMISSIVITIES

This is another case which facilitates software verification. Williams (1983, 1984) has provided an accurate (albeit non-exact) analytical closed-form solution to the case of an infinite plane slab which is homogeneous (isothermal and constant properties), grey, emitting, absorbing and isotropically scattering, and which is bounded by grey and diffuse boundaries. Williams compares his results with the accurate calculations by (Beach, Özişik and Siewert, 1971). These are regarded as benchmark standard, and are numerical, whereas Williams, whose theory is non-trivial, presents the answers in explicit closed-form solutions, in terms of tabulated functions. In the notation of Appendix D, the 1–2 view/exchange factor is

$$F_{12} = 2\hat{\Delta}^{-1}\varepsilon_1\varepsilon_2\hat{\alpha} + \hat{\Delta}^{-2}\omega\varepsilon_1\varepsilon_2\left(B_4\hat{I}_1 + B_1\hat{\alpha}\hat{I}_2\right)\left(B_4\hat{I}_1 + B_2\hat{\alpha}\hat{I}_2\right)/$$
$$\left\{0.5\omega\hat{I}_3 - \omega\hat{\Delta}^{-1}\left[\left(2 - \varepsilon_1 - \varepsilon_2\right)\hat{I}_1\hat{I}_2 + 0.5B_1B_2\hat{\alpha}\left(B_4\hat{I}_1^2 + B_3\hat{I}_2^2\right)\right]\right\} \tag{G.1}$$

where

$$\hat{I}_1 = \frac{1}{\hat{v}}\left[B_3E_2\left(\tau_L\right) - 1\right] + \frac{1}{\hat{v}^2}\left\{B_3E_1\left(\tau_L\right) - \ln\left(1 - \hat{v}\right) - E_1\left[\left(1 - \hat{v}\right)\tau_L\right]\right\} \tag{G.2}$$

$$\hat{I}_2 = \frac{1}{\hat{v}}\left[1 - B_4E_2\left(\tau_L\right)\right] + \frac{1}{\hat{v}^2}\left\{B_4E_1\left(\tau_L\right) - \ln\left(1 + \hat{v}\right) - E_1\left[\left(1 + \hat{v}\right)\tau_L\right]\right\} \tag{G.3}$$

$$\hat{I}_3 = \frac{2}{1 - \hat{v}^2} + \frac{1}{\hat{v}^2}\ln\left(1 - \hat{v}^2\right) - \frac{1}{\hat{v}\left(1 - \hat{v}\right)}E_2\left[\left(1 - \hat{v}\right)\tau_L\right] + \frac{1}{\hat{v}\left(1 + \hat{v}\right)}E_2\left[\left(1 + \hat{v}\right)\tau_L\right]$$
$$+ \frac{1}{\hat{v}^2}E_1\left[\left(1 - \hat{v}\right)\tau_L\right] + \frac{1}{\hat{v}^2}E_1\left[\left(1 + \hat{v}\right)\tau_L\right] - \frac{1}{\hat{v}^2}\left(B_3 + B_4\right)E_1\left(\tau_L\right) \tag{G.4}$$

and

$$B_3 = \exp\left(\hat{v}\tau_L\right); \quad B_4 = \exp\left(-\hat{v}\tau_L\right) \tag{G.5}$$

The $1 - g$ view factor is

$$F_{1g} = \varepsilon_1 \hat{\Delta}^{-1} \hat{\beta} \left(1 + B_2 \hat{\alpha}\right) - \varepsilon_1 \omega \hat{\Delta}^{-2} \left\{ \varepsilon_1 \left[\hat{I}_2 + B_2 B_4 \hat{\alpha} \hat{I}_1 \right] + \varepsilon_2 \left[B_4 \hat{I}_1 + B_2 \hat{\alpha} \hat{I}_2 \right] \right\}$$

$$\left\{ \varepsilon_1 \left[\hat{I}_2 + B_2 B_4 \hat{\alpha} \hat{I}_1 \right] + \varepsilon_2 \left[B_4 \hat{I}_1 + B_1 \hat{\alpha} \hat{I}_2 \right] \right\} / \left\{ \frac{\varepsilon_1}{2\hat{v}} \left(1 - B_4^2\right) + \epsilon_2 B_4 \tau_L - 0.5 \omega \hat{I}_5 \right\}$$

$$\tag{G.6}$$

$$\hat{I}_4 = \hat{v}^{-2} \left\{ \ln\left(1 + \hat{v}\right) + E_1 \left[\left(1 + \hat{v}\right)\tau_L \right] - 2 B_4 E_1 \left(\tau_L\right) + B_4^2 \left[\ln\left(1 - \hat{v}\right) + E_1 \left(\left(1 - \hat{v}\right)\tau_L\right) \right] \right\}$$

$$\tag{G.7}$$

$$\hat{I}_5 = \epsilon_1 \hat{I}_4 + B_4 \varepsilon_2 \left(\frac{2\tau_L}{\omega} - \hat{I}_3 \right) + 2 \hat{\Delta}^{-1} B_4 \hat{I}_1 \hat{I}_2 \left[\varepsilon_1 B_1 B_2 \hat{\alpha} + \varepsilon_2 \left(1 - \varepsilon_1 - \varepsilon_2\right) \right]$$

$$+ 2 \hat{\Delta}^{-1} \left(1 - \varepsilon_2\right) B_4^2 \hat{I}_1^2 \left(\varepsilon_1 + \varepsilon_2 B_1 \hat{\alpha}\right) + 2 \hat{\Delta}^{-1} \left(1 - \varepsilon_1\right) \hat{I}_2^2 \left(\varepsilon_1 + \varepsilon_2 B_2 \hat{\alpha}\right) \tag{G.8}$$

$\hat{v}$ is given by the zero of Equation (G.9) for a given ω, and varies between 0 and 1.

$$\ln\left(\frac{1 + \hat{v}}{1 - \hat{v}} \right) - \frac{2\hat{v}}{\omega} = 0 \tag{G.9}$$

Appendix H

OPTICALLY THICK LIMIT OF THE WILLIAMS THEORY FOR A GREY, ABSORBING-EMITTING AND SCATTERING SLAB

First, note that in the optically thick grey limit the net radiative flux (positive into the domain) at boundary i is

$$q_{iR} = \sigma F_{ig} \left(T_i^4 - T_g^4 \right) \tag{H.1}$$

where F_{ig} is the view factor between surface i on the one hand and the slab on the other. The notation is the same as that in Appendix D. The optically thick limit is derived from the Williams equations quoted above, and is given by the following (Sinai *et al.*, 1993):

$$F_{ig} = \left(\frac{1}{\varepsilon_i} + \frac{1}{\varepsilon_g} - 1 \right)^{-1} \tag{H.2}$$

$$\varepsilon_g = 1 - \left(\frac{2\omega}{\hat{v}^2} \right) \frac{\left[\hat{v} - \ln\left(1+\hat{v}\right) \right]^2}{\left[\hat{v} - \omega \ln\left(1+\hat{v}\right) \right]} \tag{H.3}$$

As before, ω is the scattering albedo K_s/K and $\hat{v}$ is given by the zero of Equation (G.9). These expressions attain the correct limits: If the medium is purely absorbing then the slab emissivity is 1.0, and if the medium is conservative (purely scattering) then the slab emissivity is zero. The slab emissivity, depicted in Figure H.1, is very sensitive to the albedo as the latter approaches 1.0.

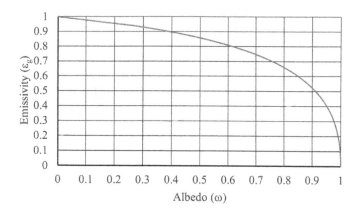

FIGURE H.1 Optically thick limit of Williams theory for emissivity of an emitting, absorbing and scattering slab.

Appendix I

INTEGRATED FORM OF THE RADIATIVE TRANSFER EQUATION

An introduction to the radiative transfer equation (RTE) has been provided in Section 4.2, and this appendix provided several 'formal' solutions to the RTE which are worth mentioning, even if their applicability is limited to a small number of issues. They are known as the *integrated forms of the RTE*. The three quoted here involve behaviour along a single ray. Consider homogeneous grey situations. For absorption only (no emission), exponential decay is predicted, known as Beer's Law, or Beer–Lambert–Bouguer Law:

$$I = I_0 \exp(-K_a s) \tag{I.1}$$

Here s is distance along the ray and I_0 is the intensity at $s = 0$. For an isothermal medium at temperature T, and in the presence of both absorption and emission, the prediction is

$$I = I_0 \exp(-K_a s) + \frac{\sigma T^4}{\pi} \left[1 - \exp(-K_a s) \right] \tag{I.2}$$

Similar relationships pertain in the spectral case; they apply at a given wavelength and with properties dependent on wavelength. As discussed previously, in a transparent medium the intensity remains constant along a give ray. If isotropic scattering is present too, then Equation (I.2) is changed to

$$I = I_0 \exp(-Ks) + \left\{ (1 - \omega) \frac{\sigma T^4}{\pi} + \omega \bar{I} \right\} \left[1 - \exp(-Ks) \right] \tag{I.3}$$

where $\bar{I}$ is the mean intensity (Equation (4.39)). The appearance of that entity reflects the complexity of scattering, since it depends on I itself and of course the intensity at all other directions too. As a first approximation in a numerical method one can assume that it is approximately constant over a computational cell during an iterative process.

Numerical approaches are common nowadays and have already been discussed.

Appendix J

SATURATED VAPOUR PRESSURE OF WATER

The Antoine correlation for the saturated vapour pressure (in Pa) is quoted here (Smith *et al.*, 2005):

$$P_{H_2Osat} = \exp\left(A - \frac{B}{T + C} \right) \tag{J.1}$$

where $A = 23.561$, $B = 4030.18$, $C = -38.15$, and T is the temperature in K. The relative humidity is the ratio of the local water vapour partial pressure to the saturated vapour pressure at the local temperature.

Appendix K

A STEADY-STATE 1-D BOUNDARY CONDITION FOR SINGLE AND DOUBLE SEMI-TRANSPARENT SLABS

This appendix summarises the results in (Sinai, 2003), which provide an idealised closed-form 1-D external boundary condition, the obvious application of which is single- and double-glazed windows. The theory may be applied to other semi-transparent materials, not only glass. The model assumes that the net volumetric radiative heating is constant across each pane, and indeed idealises the radiative aspects in what may be described as a 'lumped parameter' approach for each pane. This Appendix is particularly relevant to Section 5.1.

The model is relevant to a setup in which the domain occupies the interior of a building say, ending at the inner surface of the inner pane. A schematic is shown in Figure K.1, and is analogous to Figure 5.2, although directions have been reversed.

Since such a CFD analysis does not include the glass explicitly, the radiation can be treated on a grey basis if preferred, covering both diffuse and collimated radiation.

Since the net radiative heating per unit volume (denoted by Q_i, $i = 1, 2$) is assumed to be constant across the panes, then the solution to the conduction equation in each pane leads to a mixed boundary condition of the form shown in Equation (5.1). This follows the adoption of heat transfer correlations for the gap between the two panes (denoted by h_{CD}) and for the boundary layer at the exterior of the outer pane (E in Figure K.1), denoted by h_{EF}. The boundary condition is written here as follows:

$$\hat{A}q_B + T_B - T_S = 0; \ T_S = T_F + T_R \tag{K.1}$$

where

$$T_R = \hat{A}_1 \phi_1 + \hat{A}_2 \phi_2 \tag{K.2}$$

$$\hat{A} = \frac{L_1}{k_1} + \frac{L_2}{k_2} + \frac{1}{h_{CD}} + \frac{1}{h_{EF}} \tag{K.3}$$

$$\hat{A}_1 = \hat{A} - \frac{L_1}{2k_2}; \ \hat{A}_2 = \frac{L_2}{2k_2} + \frac{1}{h_{EF}} \tag{K.4}$$

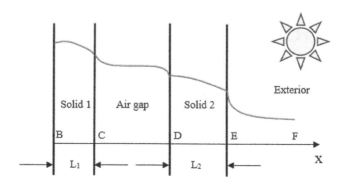

FIGURE K.1   Sketch of a 1-D double-glazing example, in which solid 1 and solid 2 are made of semi-transparent materials, and may be coated. The red curve is a schematic of the temperature profile when the interior is warmer than the exterior. The model is aimed at cases in which the computational domain is to the left of B.

$$\phi_i = Q_i L_i \left(i = 1,2 \right) \tag{K.5}$$

Here k_i is the thermal conductivity of the ith slab/pan, and ϕ_i has the meaning of net radiative heating per unit area of the given pane. The two heat transfer coefficients include convection and thermal radiation. T_F is the external or ambient temperature, i.e. the temperature well away from the assembly. T_S is a fictitious source temperature, which differs from the actual external (ambient) temperature, due to radiative heating of the panes. Sinai calls T_R the 'radiant temperature displacement'. In the absence of radiation it vanishes, as it should, and the classical results for convection and conduction emerge.

When tackling the radiant heating, the cited paper deploys further idealisations in order to retain linearity, by resorting to the macroscopic (overall) properties of each pane, embodied in the conservation of radiant energy:

$$\alpha_i + \rho_i + \tau_i = 1 \tag{K.6}$$

Here α, ρ, τ are the overall absorptivity, reflectivity and transmissivity respectively, and $i = 1$ or 2. These involve a range of aspects, such as phenomena at interfaces, multiple reflections between the panes, volumetric absorption and emission, and coatings if present. More refined analyses of the radiation are available in (Modest, 2021; Howell *et al.*, 2021). In 1-D approaches it is possible to obtain analytical or semi-analytical expressions, but the easiest approach would be to obtain experimental data. If collimated radiation is present the theory assumes it is normal to the pane, although a first estimate of effects at oblique angles might be a Beer scaling of ρ with the argument of the exponential scaling on the pane thickness multiplied by the secant of the incidence angle (relative to the normal), and in any case, the textbooks describe slab behaviour at oblique angles.

The cited paper accounts for radiation arriving at the panes from the interior too. Denoting the radiant fluxes arriving at the assembly from the exterior and interiors respectively as q_E and q_I, the heating terms are

$$\phi_1 = \frac{\alpha_1}{\left(1 - \rho_1 \rho_2\right)} \left[\tau_2 q_E + \left(1 - \rho_1 \rho_2 + \rho_2 \tau_1\right) q_I\right] \tag{K.7}$$

$$\phi_2 = \frac{\alpha_2}{\left(1 - \rho_1 \rho_2\right)} \left[\left(1 - \rho_1 \rho_2 + \rho_1 \tau_2\right) q_E + \tau_1 q_I\right] \tag{K.8}$$

These expressions acquire the correct limits if the number of panes is one ($\rho_2 = 0 = \alpha_2$, $\tau_2 = 1$), or if the panes are perfectly transparent (or in the absence of radiation), or if the panes are perfectly transparent and also possess negligible or vanishing thermal resistance.

References

Abramowitz, M. and Stegun, I. A. (1965). *Handbook of mathematical functions*. Dover Books, New York, NY.

Adiutori, E. F. (1965). 'Non-linear heat transfer phenomena', *British Chemical Engineering*, 10(12), pp. 840–845.

Al-Gebory, L. and Menguc, M. P. (2020). 'A review of optical and radiative properties of nanoparticle suspensions: effects of particle stability, agglomeration, and sedimentation', *ASME Journal of Enhanced Heat Transfer*, 27(3), pp. 207–247.

Altenhoff, M., Aßmann, S., Teige, C., Huber, F. J. T. and Will, S. (2020). 'An optimized evaluation strategy for a comprehensive morphological soot nanoparticle aggregate characterization by electron microscopy', *Journal of Aerosol Science*, 139, p. 105470.

Amaya, J., Cabrit, O., Poitou, D., Cuenot, B. and El Hafi, M. (2010). 'Unsteady coupling of Navier–Stokes and radiative heat transfer solvers applied to an anisothermal multi-component turbulent channel flow', *Journal of Quantitative Spectroscopy and Radiative Transfer*, 111(2), pp. 295–301.

Andersson, B., Andersson, R., Håkansson, L., Mortensen, M., Sudiyo, R. and Van Wachem, B. (2011). *Computational fluid dynamics for engineers*. Cambridge University Press, Cambridge, UK.

Arasteh, D., Kohler, C. and Griffith, B. (2009). *Modeling windows in energy plus with simple performance indices*. Lawrence Berkeley National Lab. (LBNL), Berkeley, CA.

Argyropoulos, C. D. and Markatos, N. C. (2015). 'Recent advances on the numerical modelling of turbulent flows', *Applied Mathematical Modelling*, 39(2), pp. 693–732.

Arking, A. (1991). 'The radiative effects of clouds and their impact on climate', *Bulletin of the American Meteorological Society*, 72(6), pp. 795–814.

Armengol, J. M., Vicquelin, R., Coussement, A., Santos, R. G. and Gicquel, O. (2020). 'Study of turbulence-radiation interactions in a heated jet using direct numerical simulation coupled to a non-gray Monte-Carlo solver', *International Journal of Heat and Mass Transfer*, 162, p. 120297.

ASHRAE (2017). *ASHRAE handbook – fundamentals*, SI Edition, Edited by M.S. Owen. ASHRAE, Atlanta, USA.

Attard, P. (2012). *Non-equilibrium thermodynamics and statistical mechanics: foundations and applications*. OUP, Oxford, UK.

Baek, S. W., Byun, D. Y. and Kang, S. J. (2000). 'The combined Monte-Carlo and finite-volume method for radiation in a two-dimensional irregular geometry', *International Journal of Heat and Mass Transfer*, 43(13), pp. 2337–2344.

Baillis, D., Raynaud, M. and Sacadura, J. F. (2000). 'Determination of spectral radiative properties of open cell foam: model validation', *Journal of Thermophysics and Heat Transfer*, 14(2), pp. 137–143.

Baker, O. (1954). 'Simultaneous flow of oil and gas', *Oil and Gas Journal*, 53, pp. 185–195.

Barnea, D. (1987). 'A unified model for predicting flow-pattern transitions for the whole range of pipe inclinations', *International Journal of Multiphase Flow*, 13(1), pp. 1–12.

Barnea, D., Shoham, O., Taitel, Y. and Dukler, A. E. (1980). 'Flow pattern transition for gas-liquid flow in horizontal and inclined pipes. Comparison of experimental data with theory', *International Journal of Multiphase Flow*, 6(3), pp. 217–225.

Batchelor, G. K. (2000). *An introduction to fluid dynamics*. Cambridge University Press, Cambridge, UK.

Baukal Jr, C. E., Gershtein, V. and Li, X. J. (2000). *Computational fluid dynamics in industrial combustion*. CRC Press, Boca Raton, FL.

Beach, H. L., Özişik, M. N. and Siewert, C. E. (1971). 'Radiative transfer in linearly anisotropic-scattering, conservative and non-conservative slabs with reflective boundaries', *International Journal of Heat and Mass Transfer*, 14(10), pp. 1551–1565.

Bird, R. B., Stewart, W. E. and Lightfoot, E. N. (2006). *Transport phenomena*, revised 2nd edn., John Wiley & Sons, New York, NY.

Blazek, J. (2015). *Computational fluid dynamics: principles and applications*. Butterworth-Heinemann, Oxford, UK.

Blunt, M. J. (2017). *Multiphase flow in permeable media: a pore-scale perspective*. Cambridge University Press, Cambridge, UK.

Böckh, P. and Wetzel, T. (2012). *Heat transfer: basics and practice*. Springer Heidelberg, Berlin, Germany.

Bohren, C. F. and Clothiaux, E. E. (2006). *Fundamentals of atmospheric radiation: an introduction with 400 problems*. John Wiley & Sons, New York, NY.

Bohren, C. F. and Huffman, D. R. (2008). *Absorption and scattering of light by small particles*. John Wiley & Sons, New York, NY.

Brantley, P. S. and Larsen, E. W. (2000). 'The simplified P 3 approximation', *Nuclear Science and Engineering*, 134(1), pp. 1–21.

Brennen, C. E. (2005). *Fundamentals of multiphase flow*, CUP, Cambridge, UK

Brewster, M. Q. (1992). *Thermal radiative transfer and properties*. John Wiley & Sons, New York, NY.

Bryner, N. P., Fuss, S. P., Klein, B. W. and Putorti Jr, A. D. (2011). *Technical Study of the Sofa Super Store Fire, South Carolina, June 18, 2007. NIST-SP 1118 Volume I*.

Buglia, J. J. (1986). *Introduction to the theory of atmospheric radiative transfer*, NASA Reference Publication 1156, Hampton, Virginia, USA.

Busbridge, I. W. (1960). *The mathematics of radiative transfer*. Cambridge University Press, Cambridge, UK.

Carlson, H., Roveda, R., Boyd, I. and Candler, G. (2004). 'A hybrid CFD-DSMC method of modeling continuum-rarefied flows', in *42nd AIAA Aerospace Sciences Meeting and Exhibit*. AIAA, p. 1180.

Casey, M. and Wintergerste, T. (2000). *ERCOFTAC special interest group on quality and trust in industrial CFD: best practice guidelines*. ERCOFTAC, Bushey, United Kingdom.

Cassol, F., Brittes, R., Centeno, F. R., da Silva, C. V. and Franca, F. H. R. (2015). 'Evaluation of the gray gas model to compute radiative transfer in nonisothermal, non-homogeneous participating medium containing CO2, H2O and soot', *Journal of the Brazilkian Society of Mechanical Sciences and Engineering*, 37(1), pp. 163–172.

Cassol, F., Brittes, R., França, F. H. R. and Ezekoye, O. A. (2014). 'Application of the weighted-sum-of-gray-gases model for media composed of arbitrary concentrations of H2O, CO2 and soot', *International Journal of Heat and Mass Transfer*, 79, pp. 796–806.

Cercignani, C. (2000). *Rarefied gas dynamics: from basic concepts to actual calculations*. Cambridge University Press, Cambridge, UK.

Chai, J. C., Lee, H. S. and Patankar, S. V. (1993). 'Ray effect and false scattering in the discrete ordinates method', *Numerical Heat Transfer, Part B Fundamentals*, 24(4), pp. 373–389.

Chandrasekhar, S. (1960). *Radiative transfer*. Dover Books, New York, NY.

Chandrasekhar, S. (2013). *Radiative transfer*. Courier Corporation, Chelmsford, MA.

Chao, Y.-A. (2016). 'A new and rigorous SPN theory for piecewise homogeneous regions', *Annals of Nuclear Energy*, 96, pp. 112–125.

Chapman, S. and Cowling, T. G. (1990). *The mathematical theory of non-uniform gases: an account of the kinetic theory of viscosity, thermal conduction and diffusion in gases*. Cambridge University Press, Cambridge.

Chatterjee, P., Wang, Y., Meredith, K. V. and Dorofeev, S. B. (2015). 'Application of a subgrid soot-radiation model in the numerical simulation of a heptane pool fire', *Proceedings of the Combustion Institute*, 35(3)., pp. 2573–2580.

Chen, J. A. and Churchill, S. W. (1963). 'Radiant heat transfer in packed beds', *AIChE Journal*, 9(1)., pp. 35–41.

Chen, J. C. (1963). *Simultaneous radiative and convective heat transfer in an absorbing, emitting, and scattering medium in slug flow between parallel plates, BNL 6876.* Brookhaven National Lab, Upton, NY.

Chen, Y. H., Bakrania, S. D., Wooldridge, M. S. and Sastry, A. M. (2010). 'Image analysis and computer simulation of nanoparticle clustering in combustion systems', *Aerosol Science and Technology*, 44(1), pp. 83–95.

Cheremisinoff, N. P. (1986). *Encyclopedia of fluid mechanics.* Edited by N. P. Cheremisinoff. Gulf, Houston, TX.

CIBSE (2000). *Testing buildings for air leakage*, CIBSE TM23, CIBSE, London, UK.

CIBSE (2021). *Environmental design, CIBSE guide A*. Edited by D. Braham. CIBSE, London, UK.

Clark, J. A. and Korybalski, M. E. (1974). 'Algebraic methods for the calculation of radiation exchange in an enclosure', *Wärme-und Stoffübertragung*, 7(1), pp. 31–44.

Clarke, J. F. and McChesney, M. (1976). *Dynamics of relaxing gases.* Butterworth, London, UK.

Clough, S. A., Shephard, M. W., Mlawer, E. J., Delamere, J. S., Iacono, M. J., Cady-Pereira, K., Boukabara, S. and Brown, P. D. (2005). 'Atmospheric radiative transfer modeling: a summary of the AER codes', *Journal of Quantitative Spectroscopy and Radiative Transfer*, 91(2), pp. 233–244.

Coelho, P. J. (2007). 'Numerical simulation of the interaction between turbulence and radiation in reactive flows', *Progress in Energy and Combustion Science*, 33(4), pp. 311–383.

Coelho, P. J. (2012). 'Turbulence–radiation interaction: from theory to application in numerical simulations', *Journal of Heat Transfer*, 134(3),031001.

Coelho, P. J. (2013). 'A theoretical analysis of the influence of turbulence on radiative emission in turbulent diffusion flames of methane', *Combustion and Flame*, 160(3), pp. 610–617.

Coelho, P. J. (2014). 'Advances in the discrete ordinates and finite volume methods for the solution of radiative heat transfer problems in participating media', *Journal of Quantitative Spectroscopy and Radiative Transfer*, 145, pp. 121–146.

Coelho, P. J. M. (2001). 'The role of ray effects and false scattering on the accuracy of the standard and modified discrete ordinates methods', in *RADIATION III. ICHMT Third International Symposium on Radiative Transfer*. Begell House Inc., Danbury, CT.

Coleman, H. W. et al. (2009). *Standard for verification and validation in computational fluid dynamics and heat transfer, VV20-2009(R2016).* ASME, New York, NY.

Coles, D. (1956). 'The law of the wake in the turbulent boundary layer', *Journal of Fluid Mechanics*, 1(2), pp. 191–226.

Coppalle, A. and Vervisch, P. (1983). 'The total emissivities of high-temperature flames', *Combustion and Flame*, 49(1–3), pp. 101–108.

Costa, M. and Azevedo, J. L. T. (2007). 'Experimental characterization of an industrial pulverized coal-fired furnace under deep staging conditions', *Combustion Science and Technology*, 179(9), pp. 1923–1935.

Cox, G. (1977). 'On radiant heat transfer from turbulent flames', *Combustion Science and Technology*, 17(1–2), pp. 75–78.

Cox, G. (1998). 'Turbulent closure and the modelling of fire by using computational fluid dynamics', *Philosophical Transactions of the Royal Society of London. Series A: Mathematical, Physical and Engineering Sciences*, 356(1748), pp. 2835–2854.

Crawley, D. B. et al. (2004). 'EnergyPlus: an update', *Proceedings of SimBuild* SimBuild 2004, IBPSA-USA National Conference, Boulder, CO, USA, 4–6 August 2004, 1, pp. 1–8.

Crosbie, A. L. and Schrenker, R. G. (1982). 'Exact expressions for radiative transfer in a three-dimensional rectangular geometry', *Journal of Quantitative Spectroscopy and Radiative Transfer*, 28(6), pp. 507–526.

Crosbie, A. L. and Schrenker, R. G. (1984). 'Radiative transfer in a two-dimensional rectangular medium exposed to diffuse radiation', *Journal of Quantitative Spectroscopy and Radiative Transfer*, 31(4), pp. 339–372.

Cushman-Roisin, B. and Beckers, J.-M. (2011). *Introduction to geophysical fluid dynamics: physical and numerical aspects*. Academic Press, Cambridge, MA.

Czerwinska, J. (2009). *Continuum and non-continuum modelling of nanofluidics, Report RTO-EN-AV-169, R&T Organization, NATO*.

Dalzell, W. H. and Sarofim, A. F. (1969). 'Optical constants of soot and their application to heat-flux calculations', *Journal of Heat Transfer*, 91(1), pp. 100–104.

Dauvois, Y., Rochais, D., Enguehard, F. and Taine, J. (2017). 'Statistical radiative modeling of a porous medium with semi transparent and transparent phases: application to a felt of overlapping fibres', *International Journal of Heat and Mass Transfer*, 106, pp. 601–618.

Davis, A. B. and Marshak, A. (2010). 'Solar radiation transport in the cloudy atmosphere: a 3D perspective on observations and climate impacts', *Reports on Progress in Physics*, 73(2), p. 26801.

Dean, R. B. (1976). 'A single formula for the complete velocity profile in a turbulent boundary layer', *Journal of Fluids Engineering*, 98(4), pp. 723–726.

Deardorff, J. W. (1980). 'Stratocumulus-capped mixed layers derived from a three-dimensional model', *Boundary-Layer Meteorology*, 18(4), pp. 495–527.

Deissler, R. G. (1964). 'Diffusion approximation for thermal radiation in gases with jump boundary condition', *Journal of Heat Transfer*, 86(2), pp. 240–246.

Demarco-Bull, R. A. (2012). *Modelling thermal radiation and soot formation in buoyant diffusion flames; Modelisation du rayonnement thermique et de la formation de suies dans des flammes de diffusion affectes par des forces de flottabilite*, PhD Thesis, Universite de Provence, France.

Denison, M. K. and Webb, B. W. (1993). 'A spectral line-based weighted-sum-of-gray-gases model for arbitrary RTE solvers', *Journal of Heat Transfer*, 115, pp. 1004–1012.

DiMarzio, C. A. (2011). *Optics for engineers*. CRC Press, Boca Raton, FL.

Divo, E. and Kassab, A. J. (2006). 'An efficient localized radial basis function meshless method for fluid flow and conjugate heat transfer', *Journal of Heat Transfer*, 129(2), pp. 124–136.

Docherty, P. (1982). 'Prediction of gas emissivity for a wide range of process conditions', 7th *International Heat Transfer Conference*, 6–10 September 1982, Munich, Vol. R5, pp. 181–185.

Dombrovsky, L. A. (2002). 'A modified differential approximation for thermal radiation of semitransparent nonisothermal particles: application to optical diagnostics of plasma spraying', *Journal of Quantitative Spectroscopy and Radiative Transfer*, 73(2–5), pp. 433–441.

Dombrovsky, L. A. and Baillis, D. (2010). *Thermal radiation in disperse systems*. Begell House, Danbury, CT.

Drikakis, D. and Frank, M. (2015). 'Advances and challenges in computational research of micro- and nanoflows', *Microfluidics and Nanofluidics*, 19(5), pp. 1019–1033.

Duo, J. I. and Azmy, Y. Y. (2007). 'Error comparison of diamond difference, nodal, and characteristic methods for solving multidimensional transport problems with the discrete ordinates approximation', *Nuclear Science and Engineering*, 156(2), pp. 139–153.

Edwards, D. K. (1976). 'Molecular gas band radiation', in T. F. Irvine, Jr. and J. P. Hartnett (eds.), *Advances in heat transfer*, 12, Elsevier, New York, NY, pp. 115–193.

Edwards, D. K. and Balakrishnan, A. (1973). 'Thermal radiation by combustion gases', *International Journal of Heat and Mass Transfer*, 16(1), pp. 25–40.

Edwards, D. K. and Menard, W. A. (1964). 'Comparison of models for correlation of total band absorption', *Applied Optics*, 3(5), pp. 621–625.

Elger, D. F., LeBret, B. A., Crowe, C. T. and Roberson, J. A. (2020). *Engineering fluid mechanics*. John Wiley & Sons, New York, NY.

Eymet, V., Brasil, A. M., El Hafi, M., Farias, T. L. and Coelho, P. J. (2002). 'Numerical investigation of the effect of soot aggregation on the radiative properties in the infrared region and radiative heat transfer', *Journal of Quantitative Spectroscopy and Radiative Transfer*, 74(6), pp. 697–718.

Faeth, G. M., Gore, J. P., Chuech, S. G. and Jeng, S.-M. (1989). 'Radiation from turbulent diffusion flames', *Annual Review of Heat Transfer*, 2, pp. 1–38.

Faeth, G. M. and Köylü, Ü. Ö. (1995). 'Soot morphology and optical properties in nonpremixed turbulent flame environments', *Combustion Science and Technology*, 108(4–6), pp. 207–229.

Farag, I. H. (1982). 'Non luminous gas radiation: approximate emissivity models', in U. Grigull, E. Hahne, K. Stephan, and J. Straub (eds.), *International heat transfer conference digital library*. Begell House Inc., Danbury, CT, pp. 487–492.

Farias, T. L., Carvalho, M. da G., Köylü, Ü. and Faeth, G. M. (1995). 'Computational evaluation of approximate Rayleigh–Debye–Gans/fractal-aggregate theory for the absorption and scattering properties of soot', *ASME Journal of Heat Transfer*, 117, pp. 152–159.

Felske, J. D. and Tien, C. L. (1973). 'Calculation of the emissivity of luminous flames', *Combustion Science and Technology*, 7(1), pp. 25–31.

Fennel, D. (1988). *Investigation into the King's cross underground fire*, Cm 499, Her Majesty's Stationery Office, London, UK.

Fernandes, C. S., Fraga, G. C., França, F. H. R. and Centeno, F. R. (2021). 'Radiative transfer calculations in fire simulations: an assessment of different gray gas models using the software FDS', *Fire Safety Journal*, 120, p. 103103.

Ferziger, J. H., Perić, M. and Street, R. L. (2002). *Computational methods for fluid dynamics*. Springer.

Fletcher, C. A. J. (1991). *Computational techniques for fluid dynamics 1*, 2nd edn., Springer-Verlag, Berlin, Germany.

Fomin, B. A. (2004). 'A k-distribution technique for radiative transfer simulation in inhomogeneous atmosphere: 1. FKDM, fast k-distribution model for the longwave', *Journal of Geophysical Research: Atmospheres*, 109(D02110), pp. 1–11.

Foster, J. A. and Roberts, G. V. (1995). 'Fire ventilation trials', *FRDG Publication Number 15/96*, Home Office Fire Research and Development Group, London, UK.

Foster, P. J. and Howarth, C. R. (1968). 'Optical constants of carbons and coals in the infrared', *Carbon*, 6(5), pp. 719–729.

Fox, R. O. (2003). *Computational models for turbulent reacting flows*. Cambridge University Press, Cambridge, UK.

Fox, R. W., McDonald, A. T. and Mitchell, J. W. (2020). *Fox and McDonald's introduction to fluid mechanics*, 10th edn., John Wiley & Sons, New York, NY.

Fraga, G. C., Centeno, F. R., Petry, A. P., Coelho, P. J. and França, F. H. R. (2019). 'On the individual importance of temperature and concentration fluctuations in the turbulence-radiation interaction in pool fires', *International Journal of Heat and Mass Transfer*, 136, pp. 1079–1089.

Fraga, G. C., Coelho, P. J., Petry, A. P. and França, F. H. R. (2020). 'Development and testing of a model for turbulence-radiation interaction effects on the radiative emission', *Journal of Quantitative Spectroscopy and Radiative Transfer*, 245, p. 106852.

Franke, J., Hellsten, A., Schlunzen, K. H. and Carissimo, B. (2011). 'The COST 732 Best Practice Guideline for CFD simulation of flows in the urban environment: a summary', *International Journal of Environment and Pollution*, 44(1–4), pp. 419–427.

Fuss, S. P. and Hamins, A. (2002). 'An estimate of the correction applied to radiant flame measurements due to attenuation by atmospheric CO_2 and H_2O', *Fire Safety Journal*, 37(2), pp. 181–190.

Garg, S. and Pant, M. (2018). 'Meshfree methods: a comprehensive review of applications', *International Journal of Computational Methods*, 15(04), p. 1830001.

Garnier, E., Adams, N. and Sagaut, P. (2009). *Large eddy simulation for compressible flows*. Springer Science & Business Media, Dordrecht, the Netherlands.

Gibbs, J. and Joyner, P. L. (1978). *Liddell boiler reheat metal temperature: gas side heat transfer analysis, Report R/M/N1029,* Central Electricity Generating Board, Leatherhead, UK.

Glatt, L. and Olfe, D. B. (1973). 'Radiative equilibrium of a gray medium in a rectangular enclosure', *Journal of Quantitative Spectroscopy and Radiative Transfer*, 13(9), pp. 881–895.

Glicksman, L. R. and Torpey, M. R. (1988). *A study of radiative heat transfer through foam insulation*. Massachusetts Inst. of Tech, Cambridge, MA.

Goldstein, R. (2017). *Fluid mechanics measurements*, 2nd edn., Routledge, Boca Raton, CA.

Goody, R. M. and Yung, Y. L. (1995). *Atmospheric radiation: theoretical basis*, 2nd edn., Oxford University Press, Oxford, UK.

Gordon, I. E. et al. (2017). 'The HITRAN2016 molecular spectroscopic database', *Journal of Quantitative Spectroscopy and Radiative Transfer*, 203, pp. 3–69.

Goutiere, V., Liu, F. and Charette, A. (2000). 'An assessment of real-gas modelling in 2D enclosures', *Journal of Quantitative Spectroscopy and Radiative Transfer*, 64(3), pp. 299–326.

Grandy Jr, W. T. and Grandy, W. T. (2005). *Scattering of waves from large spheres*. Cambridge University Press, Cambridge, UK.

Griffiths, D. J. and Schroeter, D. F. (2018). *Introduction to quantum mechanics*. Cambridge University Press, Cambridge, UK.

De Groot, S. R. and Mazur, P. (2013). *Non-equilibrium thermodynamics*. Courier Corporation, North Chelmsford, Massachusetts, USA.

Grosshandler, W. L. (1993). 'RADCAL: a narrow-band model for radiation', *Calculations in a Combustion Environment, NIST Technical Note*, 1402.

Guo, J., Shen, L., He, X., Liu, Z. and Im, H. G. (2021). 'Assessment of weighted-sum-of-gray-gases models for gas-soot mixture in jet diffusion flames', *International Journal of Heat and Mass Transfer*, 181, p. 121907.

Habibi, A., Merci, B. and Roekaerts, D. (2007). 'Turbulence radiation interaction in Reynolds-averaged Navier–Stokes simulations of nonpremixed piloted turbulent laboratory-scale flames', *Combustion and Flame*, 151(1–2), pp. 303–320.

Hanjalić, K. and Launder, B. (2011). *Modelling turbulence in engineering and the environment: second-moment routes to closure*. Cambridge University Press, Cambridge, UK.

Haupt, S. E., et al. (2019). 'On bridging a modeling scale gap: mesoscale to microscale coupling for wind energy', *Bulletin of the American Meteorological Society*, 100(12), pp. 2533–2550.

Hayt, W. H. and Buck, J. A. (2019). *Engineering electromagnetics*, 9th edn., McGraw-Hill, New York, NY.

Heald, M. A. and Marion, J. B. (2012). *Classical electromagnetic radiation*. Courier Corporation.

Heaslet, M. A. and Warming, R. F. (1965). 'Radiative transport and wall temperature slip in an absorbing planar medium', *International Journal of Heat and Mass Transfer*, 8(7), pp. 979–994.

Hecht, E. (2017). *Optics*. Pearson Education.

Hetsroni, Gad (1982). *Handbook of multiphase systems*. Edited by G. Hetsroni. McGraw-Hill Book Co., New York, NY.

Hirsch, C. (2007). *Numerical computation of internal and external flows: the fundamentals of computational fluid dynamics*, 2nd edn., Elsevier, Oxford, UK.

Hofgren, P. H. A. (2015). *Modelling and experimental investigations on thermal radiation in combustion environments*. Lund University, Sweden.

Hogan, R. J. and Shonk, J. K. P. (2013). 'Incorporating the effects of 3D radiative transfer in the presence of clouds into two-stream multilayer radiation schemes', *Journal of the Atmospheric Sciences*, 70(2), pp. 708–724.

Hollas, J. M. (2004). *Modern spectroscopy*. John Wiley & Sons, New York, NY.

Holman, J. P. (2018). *Heat transfer*, 10th edn., International Edition, McGraw-Hill.

Hosain, M. L., Domínguez, J. M., Fdhila, R. B. and Kyprianidis, K. (2019). 'Smoothed particle hydrodynamics modeling of industrial processes involving heat transfer', *Applied Energy*, 252, p. 113441.

Hostikka, S., McGrattan, K. B. and Hamins, A. (2003). 'Numerical modeling of pool fires using les and Finite Volume Method for radiation', *Fire Safety Science*, 7, pp. 383–394.

Hottel, H. C. (1954). 'Radiant heat transmission', in W. H. McAdams (ed.), *Heat transmission*, 3rd edn., McGraw-Hill Kogakusha, New York, NY.

Hottel, H. C. and Egbert, R. B. (1942). 'Radiant heat transmission from water vapor', *Transactions of the American Institute of Chemical Engineers*, 38(3), pp. 531–568.

Hottel, H. C., Noble, J. J., Sarofim, A. F., Silcox, G. D., Wankat, P. C. and Knaebel, K. S. (2007). 'Heat and mass transfer', in D. W. Green, R. H. Perry (eds.), *Perry's chemical engineers' handbook*, 8th edn., McGraw-Hill, New York, NY.

Hottel, H. C. and Sarofim, A. F. (1967). *Radiative transfer*. McGraw-Hill, New York, NY.

Howell, J. R. and Menguc, M. P. (2011). 'Radiative transfer configuration factor catalogue: a listing of relations for common geometries', *Journal of Quantitative Spectroscopy and Radiative Transfer*, 112(5), pp. 910–912.

Howell, J. R., Menguc, M. P., Daun, K. and Siegel, R. (2021). *Thermal radiation heat transfer*, 7th edn., CRC Press, Boca Raton, FL.

Hu, B. and Köylü, Ü. (2004). 'Size and morphology of soot particulates sampled from a turbulent nonpremixed acetylene flame', *Aerosol Science and Technology*, 38(10), pp. 1009–1018.

Hubbard, G. L. and Tien, C. L. (1978). 'Infrared mean absorption coefficients of luminous flames and smoke', *ASME Journal of Heat Transfer*, 100, pp. 235–239.

van de Hulst, H. C. (1981). *Light scattering by small particles*. Courier Corporation, Chelmsford, MA.

van de Hulst, H. C. (2012). *Multiple light scattering: tables, formulas, and applications*. Elsevier, Amsterdam, the Netherlands.

Hunter, B. and Guo, Z. (2015). 'Numerical smearing, ray effect, and angular false scattering in radiation transfer computation', *International Journal of Heat and Mass Transfer*, 81, pp. 63–74.

Hurley, M. J. et al. (eds) (2016). *SFPE handbook of fire protection engineering*, 5th edn., Springer, New York, NY.

Incropera, F. P., DeWitt, D. P., Bergman, T. L. and Lavine, A. S. (2017). *Incropera's principles of heat and mass transfer*. Global. Wiley, Singapore.

Ingham, D. B. and Pop, I. (eds.) (1998). *Transport phenomena in porous media*. Pergamon, Kidlington, UK.

Jacobson, M. Z. (2005). *Fundamentals of atmospheric modeling*. Cambridge University Press, Cambridge, UK.

Johnson, R. W. (ed.) (2016). *Handbook of fluid dynamics*, 2nd edn., CRC Press, Boca Raton, FL.

Joseph, D., Perez, P., El Hafi, M. and Cuenot, B. (2009). 'Discrete ordinates and Monte Carlo methods for radiative transfer simulation applied to computational fluid dynamics combustion modeling', *Journal of Heat Transfer*, 131(5), 052701 (9 pages).

Josyula, E., Xu, K. and Wadsworth, D. C. (2005). 'Testing continuum and non-continuum descriptions in high-speed flows', in *AIP Conference Proceedings*. American Institute of Physics, pp. 1217–1222.

Jullien, R. and Botet, R. (1987). *Aggregation and fractal aggregates*. World Scientific Publishing, Singapore.

Kamdem, H. T. T. (2015). 'Ray effects elimination in discrete ordinates and finite volume methods', *Journal of Thermophysics and Heat Transfer*, 29(2), pp. 306–318.

Von Karman, T. (1931). *Mechanical Similitude and Turbulence, NACA Technical Memo 611*, 1931.

Karniadakis, G., Beskok, A. and Aluru, N. (2006). *Microflows and nanoflows: fundamentals and simulation*. Springer Science & Business Media, New York, NY.

Katz, A. J. (2009). *Meshless methods for computational fluid dynamics*, PhD Thesis, Stanford University, Stanford.

Kaviany, M. (2012). *Principles of heat transfer in porous media*. Springer Science & Business Media, New York, NY.

Kavokine, N., Netz, R. R. and Bocquet, L. (2021). 'Fluids at the nanoscale: from continuum to subcontinuum transport', *Annual Review of Fluid Mechanics*, 53(1), pp. 377–410.

Kogan, M. N. (1973). 'Molecular gas dynamics', *Annual Review of Fluid Mechanics*, 5(1), pp. 383–404.

Kondrat'Yev, K. Y. (1965). *Radiative heat exchange in the atmosphere*. Translated from the second Russian edition by O. Tedder & C. D. Walshaw, Pergamon, New York, NY.

Kounalakis, M. E., Gore, J. P. and Faeth, G. M. (1989). 'Turbulence/radiation interactions in nonpremixed hydrogen/air flames', *Proc. 22nd Symposium (International) on Combustion*, 14-19 August 1988, Seattle, in *Symposium (international) on combustion*. Elsevier, 22(1), pp. 1281–1290.

Krishnamoorthy, G., Borodai, S., Rawat, R., Spinti, J. and Smith, P. J. (2005). 'Numerical modeling of radiative heat transfer in pool fire simulations', in *ASME International Mechanical Engineering Congress and Exposition*, 5–11 November 2005, Orlando, FL, USA, 42215, pp. 327–337.

Krishnamoorthy, G., Sami, M., Orsino, S., Perera, A., Shahnam, M. and Huckaby, E. D. (2010). 'Radiation modelling in oxy-fuel combustion scenarios', *International Journal of Computational Fluid Dynamics*, 24(3–4), pp. 69–82.

Krishnan, S. S., Faeth, G. M., Baum, H. R. and Officer, N. S. (2000). *Buoyant turbulent jets and flames: II. Refractive index, extinction and scattering properties of soot*. National Institute of Standards and Technology, Building and Fire Research, Gaithersburg, USA.

Ku, J. C. and Shim, K.-H. (1991). 'Optical diagnostics and radiative properties of simulated soot agglomerates', *Journal of Heat Transfer*, 113(4), pp. 953–958.

Kuerten, J. G. M. (2016). 'Point-particle DNS and LES of particle-laden turbulent flow-a state-of-the-art review', *Flow, Turbulence and Combustion*, 97(3), pp. 689–713.

Kunes, J. (2012). *Dimensionless physical quantities in science and engineering*. Elsevier, Amsterdam, the Netherlands.

Lallemant, N., Sayre, A. and Weber, R. (1996). 'Evaluation of emissivity correlations for H2O-CO2-N2/AIR mixtures and coupling with solution methods of the radiative transfer equation', *Progress in Energy and Combustion Science*, 22(6), pp. 543–574.

Lamb, H. (1994). *Hydrodynamics*, 6th edn., Cambridge University Press, Cambridge, UK.

Landau, L. D. and Lifshitz, E. M. (1987). *Fluid mechanics, volume 6 of theoretical physics*, 2nd edn., Translated by J. B. Sykes and W. H. Reid. Pergamon, Oxford, United Kingdom.

Larkin, B. K. and Churchill, S. W. (1959). 'Heat transfer by radiation through porous insulations', *AIChE Journal*, 5(4), pp. 467–474.

Leckner, B. (1972). 'Spectral and total emissivity of water vapor and carbon dioxide', *Combustion and Flame*, 19(1), pp. 33–48.

Lee, H. and Buckius, R. O. (1982). 'Scaling anisotropic scattering in radiation heat transfer for a planar medium', *ASME Journal of Heat Transfer*, 104(1), pp. 68–75.

Lee, H. and Buckius, R. O. (1983). 'Reducing scattering to nonscattering problems in radiation heat transfer', *International Journal of Heat and Mass Transfer*, 26(7), pp. 1055–1062.

Lee, S. C. (1989). 'Effect of fiber orientation on thermal radiation in fibrous media', *International Journal of Heat and Mass Transfer*, 32(2), pp. 311–319.

Lee, S.-C. and Cunnington, G. R. (2000). 'Conduction and radiation heat transfer in high-porosity fiber thermal insulation', *Journal of Thermophysics and Heat Transfer*, 14(2), pp. 121–136.

Lefebvre, A. H. (1984). 'Flame radiation in gas turbine combustion chambers', *International Journal of Heat and Mass Transfer*, 27(9), pp. 1493–1510.

Leschziner, M. A. and Drikakis, D. (2002). 'Turbulence modelling and turbulent-flow computation in aeronautics', *Aeronautical Journal*, 106, pp. 349–384.

Lesieur, M., Métais, O. and Comte, P. (2005). *Large-eddy simulations of turbulence*. Cambridge University Press, Cambridge, UK.

Lienhard IV, J. H. and Lienhard V, J. H. (2012). *A heat transfert textbook*, 4th edn., Phlogiston Press, Cambridge, UK.

Lin, C.-H., Ferng, Y.-M., Hsu, W.-S. and Pei, B.-S. (2010). 'Investigations on the characteristics of radiative heat transfer in liquid pool fires', *Fire Technology*, 46(2), pp. 321–345.

Linder, T. (2014). *Light scattering in fiber-based materials: a foundation for characterization of structural properties*. Luleå tekniska universitet.

Liou, K.-N. (2002). *An introduction to atmospheric radiation*. Academic Press, San Diego, CA.

Lipiński, W., Petrasch, J. and Haussener, S. (2010). 'Application of the spatial averaging theorem to radiative heat transfer in two-phase media', *Journal of Quantitative Spectroscopy and Radiative Transfer*, 111(1), pp. 253–258.

Liu, F., Becker, H. A. and Bindar, Y. (1998). 'A comparative study of radiative heat transfer modelling in gas-fired furnaces using the simple grey gas and the weighted-sum-of-grey-gases models', *International Journal of Heat and Mass Transfer*, 41(22), pp. 3357–3371.

Liu, L., Mishchenko, M. I. and Arnott, W. P. (2008). 'A study of radiative properties of fractal soot aggregates using the superposition T-matrix method', *Journal of Quantitative Spectroscopy and Radiative Transfer*, 109(15), pp. 2656–2663.

Lyons, P., Wong, J. and Bhandari, M., 'A comparison of window modelling methods in EnergyPlus 4.0', *SimBuild 2010, IBPSA-USA Fourth National Conference*, New York, NY, 11–13 August 2010, pp. 177–184.

Ma, J., Li, B.-W. and Howell, J. R. (2014). 'Thermal radiation heat transfer in one- and two-dimensional enclosures using the spectral collocation method with full spectrum k-distribution model', *International Journal of Heat and Mass Transfer*, 71, pp. 35–43.

Ma, Y., Varadan, V. K. and Varadan, V. V (1990). 'Enhanced absorption due to dependent scattering', *Journal of Heat Transfer*, 112(2), pp. 402–407.

Magnussen, B. F. and Hjertager, B. H. (1977). 'On mathematical modeling of turbulent combustion with special emphasis on soot formation and combustion', in *Symposium (International) on Combustion*. Elsevier, 16(1), pp. 719–729.

Mandhane, J. M., Gregory, G. A. and Aziz, K. (1974). 'A flow pattern map for gas—liquid flow in horizontal pipes', *International Journal of Multiphase Flow*, 1(4), pp. 537–553.

Markesteijn, A. P. (2011). *Connecting molecular dynamics and computational fluid dynamics*, PhD Thesis, Delft University of Technology. Available at: http://resolver.tudelft.nl/uuid:3421dcba-83e2-4454-9f2d-4738ec9acdbe.

Marschall, J. and Milos, F. S. (1997). 'The calculation of anisotropic extinction coefficients for radiation diffusion in rigid fibrous ceramic insulations', *International Journal of Heat and Mass Transfer*, 40(3), pp. 627–634.

Mathey, F., Cokljat, D., Bertoglio, J. P. and Sergent, E. (2006). 'Assessment of the vortex method for large eddy simulation inlet conditions', *Progress in Computational Fluid Dynamics, An International Journal*, 6(1–3), pp. 58–67.

Mathieu, J. and Scott, J. (2000). *An introduction to turbulent flow*. Cambridge University Press, Cambridge.

Maugendre, M., Yon, J., Coppalle, A. and Ouf, F. X. (2008). 'Measurement of the optical properties of soot particles. Application to the optical index determination taking into account their agglomerate structure', in *Eurotherm (5th European Thermal-Sciences Conference)*, Eindhoven, Netherlands (hal-01675897).

McAdams, W. H. (1954). *Heat transmission*. McGraw-Hill Kogakusha, New York, NY.

McGrattan, K., McDermott, R., Floyd, J., Hostikka, S., Forney, G. and Baum, H. (2012). 'Computational fluid dynamics modelling of fire', *International Journal of Computational Fluid Dynamics*, 26(6–8), pp. 349–361.

McGrattan, K., McDermott, R., Vanella, M., Hostikka, S. and Floyd, J. (2021). *Fire dynamics simulator technical reference guide volume 1: mathematical model*, 6th edn., NIST Special Publication 1018-1, NIST, Gaithersburg, Maryland, USA.

McKee, T. B. and Cox, S. K. (1974). 'Scattering of visible radiation by finite clouds', *Journal of Atmospheric Sciences*, 31(7), pp. 1885–1892.

Mehrota, A. K., Karan, K. and Behie, L. A. (1995). 'Estimated gas emissivities for equipment and process design', *Chemical Engineering Progress*, 91(9), pp. 70–77.

Menter, F. R., Kuntz, M. and Langtry, R. (2003). 'Ten years of industrial experience with the SST turbulence model', *Turbulence, Heat and Mass Transfer*, 4(1), pp. 625–632.

Michaelides, E., Crowe, C. T. and Schwarzkopf, J. D. (2016). *Multiphase flow handbook*. CRC Press, Boca Raton, FL.

Modak, A. T. (1979). 'Radiation from products of combustion', *Fire Safety Journal*, 1(6), pp. 339–361.

Modest, M. F. (1989). 'Modified differential approximation for radiative transfer in general three-dimensional media', *Journal of Thermophysics and Heat Transfer*, 3(3), pp. 283–288.

Modest, M. F. (1990). 'The improved differential approximation for radiative transfer in multidimensional media', *Journal of Heat Transfer*, 112(3), pp. 819–829.

Modest, M. F. (1991). 'The weighted-sum-of-gray-gases model for arbitrary solution methods in radiative transfer', *Journal of Heat Transfer*, 113(3), pp. 650–656.

Modest, M. F. (2021). *Radiative heat transfer*, 4th edn., Academic, New York, NY.

Modest, M. F., Cai, J., Ge, W. and Lee, E. (2014). 'Elliptic formulation of the simplified spherical harmonics method in radiative heat transfer', *International Journal of Heat and Mass Transfer*, 76, pp. 459–466.

Modest, M. F. and Riazzi, R. J. (2005). 'Assembly of full-spectrum k-distributions from a narrow-band database; effects of mixing gases, gases and nongray absorbing particles, and mixtures with nongray scatterers in nongray enclosures', *Journal of Quantitative Spectroscopy and Radiative Transfer*, 90(2), pp. 169–189.

Modest, M. F. and Zhang, H. (2000). 'The Full-Spectrum correlated-k distribution and its relationship to the weighted-sum-of-gray-gases method', *American Society of Mechanical Engineers, Heat Transfer Division, (Publication) HTD*, 366(January), pp. 75–84.

Modest, M. F. and Zhang, H. (2002). 'The full-spectrum correlated-k distribution for thermal radiation from molecular gas-particulate mixtures', *Journal of Heat Transfer*, 124(1), pp. 30–38.

Morrison, F. A. (2013). *An introduction to fluid mechanics*. Cambridge University Press, Cambridge.

Mulholland, G. W. (1995). 'Smoke production and properties', in P. J. DiNenno, et al. (eds.), *The SFPE handbook of fire protection engineering*, 2nd edn., pp. 2.217–2.227 SFPE, NFPA, Quincy, Massachusetts, USA.

Nakayama, Y. (2018). *Introduction to fluid mechanics*. Butterworth-Heinemann.

Nakod, P., Krishnamoorthy, G., Sami, M. and Orsino, S. (2013). 'A comparative evaluation of gray and non-gray radiation modeling strategies in oxy-coal combustion simulations', *Applied Thermal Engineering*, 54(2), pp. 422–432.

Nee, A. (2020). 'Hybrid lattice Boltzmann—Finite difference formulation for combined heat transfer problems by 3D natural convection and surface thermal radiation', *International Journal of Mechanical Sciences*, 173, p. 105447.

Oberkampf, W. L., Sindir, M. M. and Conlisk, A. T. (2002). *Guide for the verification and validation of computational fluid dynamics simulations, G-077-1998 (2002)*. AIAA, Reston, Virginia, USA. https://doi.org/10.2514/4.472855.001.

Okata, M. (2018). *A study on radiative transfer effects in 3D cloudy atmospheres using Monte Carlo numerical simulation.* PhD Thesis, University of Tokyo

Olfe, D. B. (1967). 'A modification of the differential approximation for radiative transfer', *AIAA Journal*, 5(4), pp. 638–643.

Oppenheim, A. K. (1956). 'Radiation analysis by network method', *Trans ASME*, 65(3), pp. 725–735.

Orloff, L., Modak, A. T. and Markstein, G. H. (1979). 'Radiation from smoke layers', *Proc. 17th Symposium (International) on Combustion*, 20 August 1978, in *17th Symposium (International) on Combustion*. Combustion Institute, Elsevier, 17 pp. 1029–1038.

Owens, M. P. and Sinai, Y. L. (1994). 'Comments on CFD modelling of pool fires', in *Proceedings Tenth International Heat Transfer Conference, Brighton, 14–18 August 1994.* IChemE, London, UK.

Ozisik, M. N. (1985). *Radiative transfer & interactions with conduction & convection.* Werbel & Peck, New York, USA.

Park, H. M., Ahluwalia, R. K. and Im, K. H. (1993). 'Three-dimensional radiation in absorbing-emitting-scattering media using the modified differential approximation', *International Journal of Heat and Mass Transfer*, 36(5), pp. 1181–1189.

Park, S. (2016). *Radiation transport in multiphase and spatially random media.* PhD Thesis, Imperial College. Imperial College London. Available at: https://doi.org/10.25560/45051.

Penner, S. S. (1959). *Quantitative molecular spectroscopy and gas emissivities.* Addison-Wesley, Reading, MA.

Pepper, D. W., Wang, X. and Carrington, D. B. (2013). 'A meshless method for modeling convective heat transfer', *Journal of Heat Transfer*, 135(1), 011003 (9 pages).

Pletcher, R. H., Tannehill, J. C. and Anderson, D. (2012). *Computational fluid mechanics and heat transfer.* CRC Press, Boca Raton, FL.

Poinsot, T. and Veynante, D. (2005). *Theoretical and numerical combustion,* 2nd edn., RT Edwards, Inc., Philadelphia, PA.

Poitou, D., Amaya, J., El Hafi, M. and Cuénot, B. (2012). 'Analysis of the interaction between turbulent combustion and thermal radiation using unsteady coupled LES/DOM simulations', *Combustion and Flame*, 159(4), pp. 1605–1618.

Pope, S. B. (2000). *Turbulent flows.* Cambridge University Press, Cambridge, UK.

Radel, G., Shine, K. P. and Ptashnik, I. V. (2015). 'Global radiative and climate effect of the water vapour continuum at visible and near-infrared wavelengths', *Quarterly Journal of the Royal Meteorological Society*, 141(688), pp. 727–738.

Radney, J. G. et al. (2014). 'Dependence of soot optical properties on particle morphology: measurements and model comparisons', *Environmental Science & Technology*, 48(6), pp. 3169–3176.

Raithby, G. D. and Chui, E. H. (1990). 'A finite-volume method for predicting a radiant heat transfer in enclosures with participating media', *Journal of Heat Transfer*, 112, pp. 415–423.

Randall, D. A. et al. (2007). 'Climate models and their evaluation', in Solomon, S., D. Qin, M. Manning, Z. Chen, M. Marquis, K.B. Averyt, M.Tignor and H. L. Miller (eds.), *Climate change 2007: the physical science basis. Contribution of Working Group I to the Fourth Assessment Report of the IPCC (FAR).* Cambridge University Press, Cambridge, pp. 589–662.

Ravishankar, M., Mazumder, S. and Kumar, A. (2010). 'Finite-volume formulation and solution of the p3 equations of radiative transfer on unstructured meshes', *Journal of Heat Transfer*, 132(2), 023402 (14 pages).

Reeves, D. (1956). *Flame temperature in an industrial gas turbine combustion chamber, National Gas Turbine Establishment, UK, NGTE-Memo-285.*

Roache, P. J. (1998). *Fundamentals of computational fluid dynamics.* Hermosa Publishers, lbuquerque, NM.

Rodi, W. (2017). 'Turbulence modeling and simulation in hydraulics: a historical review', *Journal of Hydraulic Engineering*, 143(5), p. 3117001.

Rodrigues, P., Gicquel, O., Franzelli, B., Darabiha, N. and Vicquelin, R. (2019). 'Analysis of radiative transfer in a turbulent sooting jet flame using a Monte Carlo method coupled to large eddy simulation', *Journal of Quantitative Spectroscopy and Radiative Transfer*, 235, pp. 187–203.

Roe, P. L. (1986). 'Characteristic-based schemes for the Euler equations', *Annual Review of Fluid Mechanics*, 18(1), pp. 337–365.

Roe, R. R. (2016). *Standard for models and simulations, NASA-STD-7009A w/CHANGE* 1.

Rohsenow, W., Hartnett, J. and Cho, Y. (1998). *Handbook of heat transfer*, 3rd edn., McGraw-Hill Education, New York, NY.

Rokhsaz, F. and Dougherty, R. L. (1989). 'Radiative transfer within a finite plane-parallel medium exhibiting fresnel reflection at a boundary', ASME HTD, *Heat Transfer Phenomena in Radiation, Combustion and Fires*, 106, pp. 1–8.

Sacadura, J.-F. (2011). 'Thermal radiative properties of complex media: theoretical prediction versus experimental identification', *Heat Transfer Engineering*, 32(9), pp. 754–770.

Sagaut, P. (2006). *Large eddy simulation for incompressible flows: an introduction*. Springer Science & Business Media, Berlin, Germany.

Sahu, D., Jain, S. and Gupta, A. (2015). 'Experimental study on methanol pool fires under low ventilated compartment', *Procedia Earth and Planetary Science*, 11, pp. 507–515.

Salby, M. L. (2012). *Physics of the atmosphere and climate*. Cambridge University Press, Cambridge, UK.

Sandia Laboratory (n.d.).. Available at: http://www.sandia.gov/TNF/radiation.html.

Scheidegger, A. E. (2020). *The physics of flow through porous media*. University of Toronto Press, Toronto, Canada.

Schlichting, H. and Gersten, K. (2017). *Boundary-layer theory*, 9th edn., Springer, Berlin, Germany.

Seung, H. and Kang, Y. H. (1999). 'Assessment of the finite-volume method and the discrete ordinates method for radiative heat transfer in a three-dimensional rectangular enclosure', *Numerical Heat Transfer: Part B: Fundamentals*, 35, pp. 85–112.

Shaddix, C. R. and Williams, T. C. (2007). 'Soot: giver and taker of light: the complex structure of soot greatly influences the optical effects seen in fires', *American Scientist*, 95(3), pp. 232–239.

Shen, C. (2010). *Rarefied gas dynamics: fundamentals, simulations and micro flows*. Springer Science & Business Media, Berlin, Germany.

Shine, K. P., Campargue, A., Mondelain, D., McPheat, R. A., Ptashnik, I. V. and Weidmann, D. (2016). 'The water vapour continuum in near-infrared windows – current understanding and prospects for its inclusion in spectroscopic databases', *Journal of Molecular Spectroscopy*, 327, pp. 193–208.

Shine, K. P., Ptashnik, I. V. and Rädel, G. (2012). 'The water vapour continuum: brief history and recent developments', *Surveys in Geophysics*, 33(3), pp. 535–555.

Siegel, R. (1973). *Net radiation method for transmission through partially transparent plates*. NASA report TN D-7384, NASA, USA.

Siegel, R. and Spuckler, C. M. (1992). 'Effect of index of refraction on radiation characteristics in a heated absorbing, emitting, and scattering layer', *ASME Transactions Journal of Heat Transfer*, 114(3), pp. 781–784.

Silva, F. R., Fraga, G. C. and Centeno, F. R. (2020). 'Comparison of GG and WSGG models for gas and soot radiation in coupled pool fire simulations', in *Proceedings 18th Brazilian Congress of Thermal Sciences and Engineering*, ABCM.

Silvestri, S., Patel, A., Roekaerts, D. J. E. M. and Pecnik, R. (2018). 'Turbulence radiation interaction in channel flow with various optical depths', *Journal of Fluid Mechanics*, 2017/11/17, 834, pp. 359–384.

Simcox, S., Wilkes, N. S. and Jones, I. P. (1988). *Fire at King's Cross Underground Station, 18th November 1987: numerical simulation of the buoyant flow and heat transfer.* Report No. AERE-G-4677, Harwell Laboratory.

Simcox, S., Wilkes, N. S. and Jones, I. P. (1992). 'Computer simulation of the flows of hot gases from the fire at King's Cross Underground station', *Fire Safety Journal*, 18(1), pp. 49–73.

Simonson, J. R. (1988). *Engineering heat transfer.* Palgrave, London, UK.

Sinai, Y. L. (1986). 'A second-order wall function for the temperature variance in turbulent flow adjacent to a diabatic wall', in *Eighth International Heat Transfer Conference, San Francisco, August 17–22, 1986.* Hemisphere, pp. 1127–1131.

Sinai, Y. L. (1987). 'A wall function for the temperature variance in turbulent flow adjacent to a diabatic wall', *ASME Journal of Heat Transfer*, 109(4), pp. 861–865.

Sinai, Y. L. (1999). 'Comments on the role of leakages in field modelling of under-ventilated compartment fires', *Fire Safety Journal*, 33(1), pp. 11–20.

Sinai, Y. L. (2003). 'A simple 1-D boundary condition representing combined convective-radiative heat transfer across a double membrane semi-transparent assembly', *International Journal of Ventilation*, 2(2), pp. 125–133.

Sinai, Y. L., Ford, I. J., Barrett, J. C. and Clement, C. F. (1993). 'Prediction of coupled heat and mass transfer in the fast reactor cover gas: the C-GAS code', *Nuclear Engineering and Design*, 140(2), pp. 159–192.

Sinai, Y. L., Noakes, C., Awbi, H., Mustakallio, P., Waterson, N. and Cook, M. (2016). *Modelling of ventilation for thermal comfort and indoor air quality, ESDU TM 180.* IHS Markit, London, UK.

Sinai, Y. L. and Owens, M. P. (1995). 'Validation of CFD modelling of unconfined pool fires with cross-wind: flame geometry', *Fire Safety Journal*, 24(1), pp. 1–34.

Sinai, Y. L., Staples, C., Edwards, M. and Smerdon, M. (2008). 'Modelling of fire and fire suppression by water spray in the marine environment using the CFX software', in *Proc. Marine CFD 2008*, 26–27 March 2008, Southampton, UK. Royal Institution of Naval Architects, London.

Singh, B. P. and Kaviany, M. (1992). 'Modelling radiative heat transfer in packed beds', *International Journal of Heat and Mass Transfer*, 35(6), pp. 1397–1405.

Singh, R. and Mishra, S. C. (2007). 'Analysis of radiative heat transfer in a planar participating medium subjected to diffuse and/or collimated radiation—a comparison of the DTM, the DOM, and the FVM', *Numerical Heat Transfer, Part A: Applications*, 52(5), pp. 481–496.

Singham, J. R. (1962). 'Tables of emissivity of surfaces', *International Journal of Heat and Mass Transfer*, 5(1–2), pp. 67–76.

Smith, J. M., Van Ness, H. C. and Abbott, M. M. (2005). *Introduction to chemical engineering thermodynamics*, (2001) and 7th edn., (2005). McGraw-Hill, New York, NY.

Smith, T. F., Shen, Z. F. and Friedman, J. N. (1982). 'Evaluation of coefficients for the weighted sum of gray gases model', *ASME Journal of Heat Transfer*, 104, pp. 602–608.

Snegirev, A. Y. (2004). 'Statistical modeling of thermal radiation transfer in buoyant turbulent diffusion flames', *Combustion and Flame*, 136(1–2), pp. 51–71.

Snelling, D. R., Link, O., Thomson, K. A. and Smallwood, G. J. (2011). 'Measurement of soot morphology by integrated LII and elastic light scattering', *Applied Physics B*, 104(2), pp. 385–397.

Sorensen, C. M. and Feke, G. D. (1996). 'The morphology of macroscopic soot', *Aerosol Science and Technology*, 25(3), pp. 328–337.

Soufiani, A. and Taine, J. (1997). 'High temperature gas radiative property parameters of statistical narrow-band model for H2O, CO2 and CO, and correlated-K model for H2O and CO2', *International Journal of Heat and Mass Transfer*, 40(4), pp. 987–991.

Spalding, D. B. (1961). 'A single formula for the "law of the wall"', *Journal of Applied Mechanics*, 28(3), 455–458.

Sparrow, E. M. and Cess, R. D. (1978). *Radiation heat transfer.* Hemisphere, Washington, DC.

Spiegel, E. A. (1957). 'The smoothing of temperature fluctuations by radiative transfer', *The Astrophysical Journal*, 126, p. 202.

Stamnes, K., Thomas, G. E. and Stamnes, J. J. (2017). *Radiative transfer in the atmosphere and ocean*, 2nd edn., Cambridge University Press, Cambridge, UK.

Stasick, J. (1988). 'Generalised net-radiation method for open and closed enclosures filled with isothermal optical medium', *Wärme – und Stoffübertragung*, 22, pp. 259–267.

Steinfeld, J. I. (2012). *Molecules and radiation: an introduction to modern molecular spectroscopy*. Courier Corporation, Chelmsford, MA.

Steward, F. R. and Kocaefe, Y. S. (1986). 'Total emissivity and absorptivity models for carbon dioxide, water vapor and their mixtures', in *International Heat Transfer Conference Digital Library*. Begell House Inc, Danbury, CT.

Streeter, V. L. and Wylie, E. B. (1983). *Fluid mechanics*. SI Metric (ed.). McGraw-Hill, New York, NY.

Suryadharma, R. N. S. (2020). *T-matrix method for the analysis of electromagnetic scattering*. PhD Thesis, Karlsruher Institut für Technologie (KIT), Karlsruhe, Germany.

Swafford, T. W. (1983). 'Analytical approximation of two-dimensional separated turbulent boundary-layer velocity profiles', *AIAA Journal*, 21(6), pp. 923–926.

Tabor, D. (1991). *Gases, liquids and solids: and other states of matter*, 3rd edn., Cambridge University Press, Cambridge, UK.

Taitel, Y. and Dukler, A. E. (1976). 'A model for predicting flow regime transitions in horizontal and near horizontal gas-liquid flow', *AIChE Journal*, 22(1), pp. 47–55.

Tam, W. C. and Yuen, W. W. (2019). *Open SC: an open-source calculation tool for combustion mixture Emissivity/absorptivity*. US Department of Commerce, National Institute of Standards and Technology, Gaithersburg, USA.

Tao, W. Q. and Sparrow, E. M. (1985). 'Ambiguities related to the calculation of radiant heat exchange between a pair', *International Journal of Heat and Mass Transfer*, 28(9), pp. 1786–1787.

Tavoularis, S. (2005). *Measurement in fluid mechanics*. Cambridge University Press, Cambridge, UK.

Taylor, P. B. and Foster, P. J. (1974). 'The total emissivities of luminous and non-luminous flames', *International Journal of Heat and Mass Transfer*, 17(12), pp. 1591–1605.

Tencer, J. (2014). 'The impact of reference frame orientation on discrete ordinates solutions in the presence of ray effects and a related mitigation technique', in *ASME International Mechanical Engineering Congress and Exposition*. American Society of Mechanical Engineers, p. V08AT10A017.

Tencer, J. and Howell, J. R. (2016). 'Coupling radiative heat transfer in participating media with other heat transfer modes', *Journal of the Brazilian Society of Mechanical Sciences and Engineering*, 38(5), pp. 1473–1487.

Tennekes, H. and Lumley, J. L. (1972). *A first course in turbulence*. MIT Press, Cambridge, MA.

Tennekes, H. and Lumley, J. L. (2018). *A first course in turbulence*. MIT Press, Cambridge, MA.

Tien, C. L. (1968). 'Thermal radiation properties of gases', *Advances in Heat Transfer*, 5, pp. 253–324.

Tien, C.-L. and Drolen, B. L. (1987). 'Thermal radiation in particulate media with dependent and independent scattering', *Annual Review of Heat Transfer*, 1, pp. 1–32.

Tien, C. L. and Lee, S. C. (1982). 'Flame radiation', *Progress in Energy and Combustion Science*, 8(1), pp. 41–59.

Tolstoy, I. (1973). *Wave propagation*. McGraw-Hill, New York, NY.

Tong, T. W. and Tien, C. L. (1980). 'Resistance-network representation of radiative heat transfer with particulate scattering', *Journal of Quantitative Spectroscopy and Radiative Transfer*, 24(6), pp. 491–503.

Torres-Monclard, K., Gicquel, O. and Vicquelin, R. (2021). 'A Quasi-Monte Carlo method to compute scattering effects in radiative heat transfer: application to a sooted jet flame', *International Journal of Heat and Mass Transfer*, 168, p. 120915.

Townsend, A. A. (1958). 'The effects of radiative transfer on turbulent flow of a stratified fluid', *Journal of Fluid Mechanics*, 4(4), pp. 361–375.

Townsend, A. A. R. (1980). *The structure of turbulent shear flow*. Cambridge University Press, Cambridge, UK.

Travkin, V. S. and Catton, I. (1999). 'Radiation heat transport in porous media', *ASME-Publications-HTD*, 364, pp. 31–40.

Tritton, D. J. (2012). *Physical fluid dynamics*. Springer, Dordrecht, the Netherlands.

Truelove, J. S. (1976). *A mixed grey gas model for flame radiation, United Kingdom Atomic Energy Authority report*, AERE-R8494, Harwell, England.

Truelove, J. S. (1984). 'Radiant heat transfer through the cover gas of a sodium-cooled fast reactor', *International Journal of Heat and Mass Transfer*, 27(11), pp. 2085–2093.

Tu, J., Yeoh, G. H. and Liu, C. (2018). *Computational fluid dynamics: a practical approach*. Butterworth-Heinemann, Oxford, UK.

Tu, R., Fang, J., Zhang, Y.-M., Zhang, J. and Zeng, Y. (2013). 'Effects of low air pressure on radiation-controlled rectangular ethanol and n-heptane pool fires', *Proceedings of the Combustion Institute*, 34(2), pp. 2591–2598.

Turner, J. S. (2012). *Buoyancy effects in fluids*, Cambridge University Press, Cambridge, UK.

Vadász, P. (2008). *Emerging topics in heat and mass transfer in porous media: from bioengineering and microelectronics to nanotechnology*. Springer, Dordrecht, ISBN 978-1-4020-8177-4.

Veerman, M. A., Pincus, R., Stoffer, R., van Leeuwen, C. M., Podareanu, D. and van Heerwaarden, C. C. (2021). 'Predicting atmospheric optical properties for radiative transfer computations using neural networks', *Philosophical Transactions of the Royal Society A*, 379, p. 20200095. Available at: https://doi.org/10.1098/rsta.2020.0095.

Versteeg, H. K. and Malalasekera, W. (2007). *An introduction to computational fluid dynamics: the finite volume method*. Pearson Education, London.

Vicquelin, R., Zhang, Y., Gicquel, O. and Taine, J. (2014). 'Effects of radiation in turbulent channel flow: analysis of coupled direct numerical simulations', *Journal of Fluid Mechanics*, 753, pp. 360–401.

Viskanta, R. (2008). 'Computation of radiative transfer in combustion systems', *International Journal of Numerical Methods for Heat & Fluid Flow*, 18(3/4), p. 415.

Viskanta, R. and Menguc, M. P. (1989). 'Radiative transfer in dispersed media', *Applied Mechanics Reviews*, 42(9), pp. 241–259.

Walters, D., et al. (2019). *The Met Office Unified Model Global Atmosphere 7.0/7.1 and JULES Global Land 7.0 configurations, Geoscientific Model Development, 12, 1909–1963*, Available at: https://doi.org/10.5194/gmd-12-1909-2019.

Wang, C., Ge, W., Modest, M. F. and He, B. (2016). 'A full-spectrum k-distribution look-up table for radiative transfer in nonhomogeneous gaseous media', *Journal of Quantitative Spectroscopy and Radiative Transfer*, 168, pp. 46–56.

Welch, R. M., Cox, S. K. and Davis, J. M. (1980). 'Solar radiation and clouds', *Meteorological Monographs*, 17(39), American Meteorological Society, London ISBN 9780933876491.

Wendisch, M. and Yang, P. (2012). *Theory of atmospheric radiative transfer: a comprehensive introduction*. John Wiley & Sons, New York, NY.

Whitfield, D. L. (1977). *Analytical description of the complete two-dimensional turbulent boundary-layer velocity profile, AEDC-TR-77-79*. Arnold Engineering Development Center Arnold AFB TN.

Wielicki, B. A., Cess, R. D., King, M. D., Randall, D. A. and Harrison, E. F. (1995). 'Mission to planet Earth: role of clouds and radiation in climate', *Bulletin of the American Meteorological Society*, pp. 2125–2153.

Williams, M. M. R. (1983). 'Radiant heat transfer through an aerosol suspended in a transparent gas', *IMA Journal of Applied Mathematics*, 31(1), pp. 37–50.

Williams, M. M. R. (1984). 'Radiant heat transfer through an aerosol suspended in a transparent gas: addendum', *IMA Journal of Applied Mathematics*, 33(1), pp. 101–103.

Wu, C. Y., Sutton, W. H. and Love, T. J. (1987). 'Successive improvement of the modified differential approximation inradiative heat transfer', *Journal of Thermophysics and Heat Transfer*, 1(4), pp. 296–300.

Yadigaroglu, G. and Hewitt, G. F. (2017). *Introduction to multiphase flow: basic concepts, applications and modelling*. Springer, Cham, Switzerland.

Yan, Z. and Holmstedt, G. (1997). 'Fast, narrow-band computer model for radiation calculations', *Numerical Heat Transfer*, 31(1), pp. 61–71.

Yang, Z. and Gopan, A. (2021). 'Improved global model for predicting gas radiative properties over a wide range of conditions', *Thermal Science and Engineering Progress*, 22, p. 100856.

Young, D. F., Munson, B. R., Okiishi, T. H. and Huebsch, W. W. (2010). *A brief introduction to fluid mechanics*. John Wiley & Sons, New York, NY.

Yuen, W. W. (1990). 'Development of a network analogy and evaluation of mean beam lengths for multidimensional absorbing/isotropically scattering media', *Journal of Heat Transfer*, 112(2), pp. 408–414.

Zeghondy, B., Iacona, E. and Taine, J. (2006). 'Determination of the anisotropic radiative properties of a porous material by radiative distribution function identification (RDFI)', *International Journal of Heat and Mass Transfer*, 49(17–18), pp. 2810–2819.

Zenier, F., Antonello, F., Dattilo, F. and Rosa, L. (2001). 'Investigation of an LPG accident with different mathematical model applications', *International Journal of Risk Assessment and Management*, 2(3–4), pp. 340–351.

Zhang, L., Xu, R. and Jiang, P. (2014). 'Comparison of volume-average simulation and pore-scale simulation of thermal radiation and natural convection in high temperature packed beds', *Proceedings of the 5th International Conference on Porous Media and its Appliations in Science and Engineering (ICPM5)*, 22–27 June 2014, Kona, Hawaii, ECI Digital Archives.

Zhang, Y. F., Vicquelin, R., Gicquel, O. and Taine, J. (2013). 'Physical study of radiation effects on the boundary layer structure in a turbulent channel flow', *International Journal of Heat and Mass Transfer*, 61, pp. 654–666.

Index

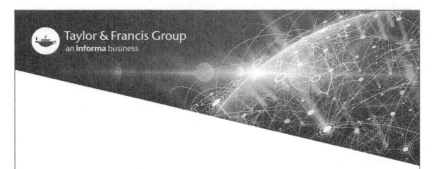